COLOR, HEALTHCARE AND BIOETHICS

Color, Healthcare and Bioethics

Henk ten Have

https://www.openbookpublishers.com
©2025 Henk ten Have

This work is licensed under a Creative Commons Attribution-NonCommercial 4.0 International (CC BY-NC 4.0). This license allows you to share, copy, distribute and transmit the text; to adapt the text for non-commercial purposes of the text providing attribution is made to the authors (but not in any way that suggests that they endorse you or your use of the work). Attribution should include the following information:

Henk ten Have, *Color, Healthcare and Bioethics*. Cambridge, UK: Open Book Publishers, 2025, https://doi.org/10.11647/OBP.0443

Further details about CC BY-NC licenses are available at http://creativecommons.org/licenses/by-nc/4.0/

All external links were active at the time of publication unless otherwise stated and have been archived via the Internet Archive Wayback Machine at https://archive.org/web

Digital material and resources associated with this volume are available at https://doi.org/10.11647/OBP.0443#resources

Information about any revised edition of this work will be provided at https://doi.org/10.11647/OBP.0443

ISBN Paperback 978-1-80511-482-6
ISBN Hardback 978-1-80511-483-3
ISBN PDF 978-1-80511-484-0
ISBN HTML 978-1-80511-486-4
ISBN EPUB 978-1-80511-485-7
DOI: 10.11647/OBP.0443

Cover image: Textures & Patterns (2022), https://unsplash.com/photos/modern-multi-colored-template-with-curve-dynamic-fluid-flow-abstract-geometric-background-3d-rendering-digital-illustration-Txu4PTomSrY
Cover design: Jeevanjot Kaur Nagpal

For Cecilia N. G. Jansen,

making our life so colorful

Contents

About the Author	xi
List of Illustrations	xiii
1. Introduction: Color, Healthcare and Bioethics	1
1.1 The Experience of Color	1
1.2 Color and Medicine	3
1.3 Color and Ethics	7
1.4 The Outline of the Book	10
1.5 The Nature of Color	10
1.6 The Power of Color	12
1.7 Color and Healthcare	15
1.8 Color and Bioethics	17
1.9 A Colorful Bioethics	21
1.10 Conclusion	24
References	25
2. The Nature of Color	29
2.1 Introduction	29
2.2 The Traditional View	31
2.3 The Scientific Revolution	31
2.4 Primary and Secondary Qualities	33
2.5 Philosophies of Color	34
2.6 Color Relationism	36
2.7 Ecological Theories	38
2.8 Color Adverbialism	40
2.9 The Phenomenological Perspective	41
2.10 Conclusion	43
References	44

3. The Power of Color	47
3.1 Introduction	47
3.2 Color Language	50
3.3 The Affective Power of Color	54
3.4 The Meaning of Color	57
3.5 Color, Emotions and Feelings	64
3.6 Color, Human Behavior and Performance	66
3.7 Practical Implications	68
3.8 Conclusion	70
References	73
4. Color and Healthcare	79
4.1 Introduction	79
4.2 Disease	81
4.3 Diagnosis	82
4.4 The Color of Medication	90
4.5 Pigments as Pharmaceuticals	92
4.6 The Pharmaceutical Revolution	96
4.7 Color Therapy	101
4.8 Healing Environments	104
4.9 Conclusion	106
References	107
5. Color and Bioethics	111
5.1 Introduction	111
5.2 Colors and Normativity	114
5.3 Color and Rationality	117
5.4 Moral Associations of Black and White	120
5.5 Color and Race	123
5.6 Skin Color	127
5.7 Color-Based Hierarchies	129
5.8 Racial Science	131
5.9 The Persistence of Race and Racism	134
5.10 Racism and Healthcare	136
5.11 Racism and Bioethics	141
5.12 The Whiteness of Bioethics	143

5.13 Conclusion	146
References	150
6. A Colorful Bioethics	159
6.1 Introduction	159
6.2 Race as Bioethical Issue	160
6.3 Race under the Bioethical Microscope	164
6.4 Racism as Bioethical Issue	172
6.5 The Color of Bioethics	178
6.7 Ethics and Aesthetics	181
6.8 Moral Imagination	190
6.9 Expansion of Bioethical Discourse	194
6.10 Conclusion	203
References	206
Index	215

About the Author

Henk ten Have has been Director of the Center for Healthcare Ethics at Duquesne University in Pittsburgh, USA (2010–2019). He studied medicine and philosophy in the Netherlands and worked as professor in the Faculty of Medicine of the Universities of Maastricht and Nijmegen. From 2003 until 2010 he has joined UNESCO in Paris as Director of the Division of Ethics of Science and Technology. Since 2019 he is Emeritus Professor, Duquesne University, Pittsburgh, USA, and since 2021 Research Professor at the Faculty of Bioethics in the Universidad Anahuac Mexico. He is editor of the *International Journal of Ethics Education*, and *Medicine, Health Care and Philosophy*.

His recent book publications are *Global Bioethics; An Introduction* (2016), *Vulnerability: Challenging Bioethics* (2016), *Global Education in Bioethics* (2018), *Wounded Planet. How Declining Biodiversity Endangers Health and How Bioethics Can Help* (2019), *Dictionary of Global Bioethics* (with Maria do Céu Patrão Neves, 2021), *Bioethics, Healthcare and the Soul* (with Renzo Pegoraro, 2022), *Bizarre Bioethics—Ghosts, Monsters and Pilgrims* (2022) and *The Covid-19 Pandemic and Global Bioethics* (2022). He has edited the *Encyclopedia of Global Bioethics* (2016) and *Global Education in Bioethics* (2018).

List of Illustrations

Fig. 1.1	Hazy blue hour in Grand Canyon, photo by Michael Gäbler (1988). Wikimedia, https://commons.wikimedia.org/wiki/File:Hazy_blue_hour_in_Grand_Canyon.JPG#/media/File:Hazy_blue_hour_in_Grand_Canyon.JPG, CC BY 3.0.	p. 2
Fig. 1.2	The four humors, Tom Lemmens (2013). Wikimedia, https://commons.wikimedia.org/wiki/File:Humorism.svg#/media/File:Humorism.svg, CC0 1.0.	p. 4
Fig. 1.3	Angelica Dass Retratos, *Proyecto Humanae Valencia*. Color labels like black, white, yellow or red are inadequate to capture this diversity. Photo by LOLAOMI (2014), Wikimedia, https://commons.wikimedia.org/wiki/File:Ang%C3%A9lica_Dass_retratos.jpg#/media/File:Ang%C3%A9lica_Dass_retratos.jpg, CC0 1.0.	p. 20
Fig. 2.1	Isaac Newton's prism experiment. Image created by Castellsferran (2020), Wikimedia, https://commons.wikimedia.org/wiki/File:Experiment_dels_primes_d'Isaac_Newton_-_Refracci%C3%B3_de_la_llum.png#/media/File:Experiment_dels_primes_d'Isaac_Newton_-_Refracci%C3%B3_de_la_llum.png, CC BY-SA 4.0.	p. 32
Fig. 2.2	Color wheel wavelengths. Image created by Amousey (2023), Wikimedia, https://commons.wikimedia.org/wiki/File:Color_wheel_vector.svg#/media/File:Color_wheel_vector.svg, CC BY-SA 4.0.	p. 33
Fig. 2.3	Yellow-banded poison dart frog. Photo by Holger Krisp (2013), Wikimedia, https://commons.wikimedia.org/wiki/File:Bumblebee_Poison_Frog_Dendrobates_leucomelas.jpg#/media/File:Bumblebee_Poison_Frog_Dendrobates_leucomelas.jpg, CC BY 3.0.	p. 39

Fig. 3.1	Joseph Karl Stieler, *Johann Wolfgang von Goethe* (1828). Neue Pinakothek, Munich. Photo by Pierre André (2016), Wikimedia, https://commons.wikimedia.org/wiki/File:Joseph_Karl_Stieler_portrait_de_Johann_Wolfgang_von_Goethe.jpg#/media/File:Joseph_Karl_Stieler_portrait_de_Johann_Wolfgang_von_Goethe.jpg, CC BY-SA 4.0.	p. 54
Fig. 3.2	William Turner, *Light and Colour (Goethe's Theory)* (1843). Tate Britain, London. Photo by Wuselig (2020), Wikimedia, https://commons.wikimedia.org/wiki/File:Horror_und_Delight-Turner-Light_and_Colour_(Goethe%27s_Theory)_DSC2252.jpg#/media/File:Horror_und_Delight-Turner-Light_and_Colour_(Goethe's_Theory)_DSC2252.jpg, CC0 1.0.	p. 57
Fig. 3.3	Utigawa Kuniyoshi, *Medical and Surgical Treatments for a Lame Princess and Others* (1849/52). Wellcome Collection. Wikimedia, https://commons.wikimedia.org/wiki/File:Medical_and_surgical_treatments_for_a_lame_princess_Wellcome_L0035015.jpg#/media/File:Medical_and_surgical_treatments_for_a_lame_princess_Wellcome_L0035015.jpg, CC BY 4.0.	p. 58
Fig. 3.4	Yellow emperor. Scan from *Shèhuì Lìshǐ Bówùguǎn* [Social History Museum]. Wikimedia, https://commons.wikimedia.org/wiki/File:Yellow_Emperor.jpg#/media/File:Yellow_Emperor.jpg, public domain.	p. 61
Fig. 4.1	Johannes de Ketham, *Fasciculus Medicinae* (1491). Uroscopy chart relating the color of urine to bodily constitutions and ailments. Wikimedia, https://commons.wikimedia.org/wiki/File:Fasciculus_Medicinae_1491.jpg#/media/File:Fasciculus_Medicinae_1491.jpg, public domain.	p. 85
Fig. 4.2	Red litmus paper reacts with hydrochloric acid in litmus test. Photo by Kanesskong (2016), Wikimedia, https://commons.wikimedia.org/wiki/File:The_result_of_red_litmus_paper.jpg#/media/File:The_result_of_red_litmus_paper.jpg, CC BY-SA 4.0.	p. 87
Fig. 4.3	Microscopic image of a Gram stain of mixed Gram-positive Staphylococcus aureus (purple) and Gram-negative Escherichia coli (red). Image by Y tambe (2010), Wikimedia, https://commons.wikimedia.org/wiki/File:Gram_stain_01.jpg#/media/File:Gram_stain_01.jpg, CC BY-SA 3.0.	p. 89

Fig. 4.4	Jerry Allison, *William Henry Perkin—Pioneer in Synthetic Organic Dyes* (1980). Science History Institute. Perkin (center) in his laboratory examines test dying of silk taffeta with mauve aniline dye. Wikimedia, https://commons.wikimedia.org/wiki/File:William_Henry_Perkin-_Pioneer_in_Synthetic_Organic_Dyes_-_DPLA_-_3acf6c4043b0ea3ee1044c835092c5ec.jpg#/media/File:William_Henry_Perkin-_Pioneer_in_Synthetic_Organic_Dyes_-_DPLA_-_3acf6c4043b0ea3ee1044c835092c5ec.jpg, CC BY 4.0.	p. 96
Fig. 4.5	Paul Ehrlich, c. 1910. Photographer unknown. Wikimedia, https://commons.wikimedia.org/wiki/File:Paul_Ehrlich,_c._1910.jpg#/media/File:Paul_Ehrlich,_c._1910.jpg, public domain.	p. 99
Fig. 4.6	Phototherapy of neonate for jaundice. Photo by Vtbijoy (2013), Wikimedia, https://commons.wikimedia.org/wiki/File:Phototherapy.jpg#/media/File:Phototherapy.jpg, CC BY-SA 3.0.	p.103
Fig. 5.1	Cistercian monks. Bernard of Clairvaux invests Gerwig with the robes of the Cistercian order. Fresco from 1695–1698 by Johann Jakob Steinfels in Abbey church Waldsassen. Photo by Wolfgang Sauber (2018), Wikimedia, https://commons.wikimedia.org/wiki/File:Waldsassen_Stiftsbasilika_-_Fresko_3c_Gr%C3%BCndungslegende.jpg#/media/File:Waldsassen_Stiftsbasilika_-_Fresko_3c_Gr%C3%BCndungslegende.jpg, CC BY-SA 4.0.	p.115
Fig. 5.2	Rembrandt, *The Anatomy Lesson of Dr. Nicolaes Tulp* (1632). Mauritshuis, The Hague. Wikimedia, https://commons.wikimedia.org/wiki/File:Rembrandt_-_The_Anatomy_Lesson_of_Dr_Nicolaes_Tulp.jpg#/media/File:Rembrandt_-_The_Anatomy_Lesson_of_Dr_Nicolaes_Tulp.jpg, public domain.	p.119
Fig. 5.3	Jean-Auguste-Dominique Ingres, *Portrait of Francois Bernier* (1800). Wikimedia, https://commons.wikimedia.org/wiki/File:Bernier-Ingres-1800.jpg#/media/File:Bernier-Ingres-1800.jpg, public domain.	p.124

Fig. 5.4	Johann Friedrich Blumenbach, *De generis humani varietate* (1795). Sequence of human skulls showing the diversity of the main types. Wellcome Collection. Wikimedia, https://commons.wikimedia.org/wiki/File:J.F._Blumenbach,_De_generis_humani_varietate_Wellcome_L0032295.jpg#/media/File:J.F._Blumenbach,_De_generis_humani_varietate_Wellcome_L0032295.jpg, CC BY 4.0.	p.126
Fig. 5.5	"Colored" water cooler in streetcar terminal in Oklahoma City (1939). Wikimedia, https://commons.wikimedia.org/wiki/File:%22Colored%22_drinking_fountain_from_mid-20th_century_with_african-american_drinking.jpg, public domain.	p.140
Fig. 6.1	Caucasian Biosphere Reserve in the vicinities of Sochi, Russian Federation. Photo by SKas (2016), Wikimedia, https://commons.wikimedia.org/wiki/File:Caucasian_Biosphere_Reserve.jpg#/media/File:Caucasian_Biosphere_Reserve.jpg, CC BY-SA 4.0.	p.165
Fig. 6.2	Alex da Silva, *Slavery Monument* (2013), Rotterdam. Photo by GraphyArchy (2020), Wikimedia, https://commons.wikimedia.org/wiki/File:GraphyArchy_-_Wikipedia_00706.jpg#/media/File:GraphyArchy_-_Wikipedia_00706.jpg, CC BY-SA 4.0.	p.176
Fig. 6.3	Skin-whitening product in supermarket in Sri Lanka. Photo by Adam Jones (2014), Wikimedia, https://commons.wikimedia.org/wiki/File:Fair_and_Handsome_-_Skin-Whitening_Product_in_Supermarket_-_Bandarawela_-_Hill_Country_-_Sri_Lanka_(14122094934).jpg#/media/File:Fair_and_Handsome_-_Skin-Whitening_Product_in_Supermarket_-_Bandarawela_-_Hill_Country_-_Sri_Lanka_(14122094934).jpg, CC BY-SA 2.0.	p.184
Fig. 6.4	Gaston Bachelard (1965), Dutch National Archives, The Hague. Photographer unknown, uploaded by Anefo, Wikimedia, https://commons.wikimedia.org/wiki/File:Gaston_Bachelard_(kop)_filosoof,_Bestanddeelnr_917-9599.jpg#/media/File:Gaston_Bachelard_(kop)_filosoof,_Bestanddeelnr_917-9599.jpg, CC0.	p.191

1. Introduction:
Color, Healthcare and Bioethics

1.1 The Experience of Color

In "L'heure bleue" (1970) the French artist Françoise Hardy sings about the brief moment when the day has ended but the night has not yet started; an uncertain juncture where everything becomes more beautiful, softer and brighter. It is also a happy time where you wait for the person you love. Listening to this song, I was wondering why this brief period of twilight is called the "blue" hour. The usual explanation is that when the sun is sinking below the horizon, the shorter wavelengths of the visible light dominate so that the remaining light has a blue shade. The expression is not very common in the Dutch language, perhaps because the skies are often more cloudy than in the southern parts of Europe. But a related expression is "a blue Monday" referring to an extremely short period of time in which people have worked or lived in a particular place. The blue hour has inspired numerous painters, musicians and writers, not because it is a meteorological condition or a specific time of the day, but since it has manifold symbolic meanings. As the transition between light and dark, it provides the occasion to dream, and to reflect on what has ended and what will be expected. It is a melancholic experience, an awareness of loss and impermanence. The blue hour is the point of metamorphosis; it provokes not only mourning about what is over, but also longing for restoration and renewal. As an occasion of hope, it connects nostalgia and sadness with the expectation of happiness and new beginnings.

Fig. 1.1 Hazy blue hour in Grand Canyon, photo by Michael Gäbler (1988). Wikimedia, https://commons.wikimedia.org/wiki/File:Hazy_blue_hour_in_Grand_Canyon.JPG#/media/File:Hazy_blue_hour_in_Grand_Canyon.JPG, CC BY 3.0.

This example illustrates that color is a physical and physiological phenomenon which can be explained in terms of different wavelengths of light, and properties of our visual system. During the twilight period the sky is dark blue, and we can observe this color, while the sciences of physics and physiology help us to explain why this occurs. At the same time, the blue hour is an experience associated with emotions and feelings. The scientific explanation is not sufficient to comprehend the phenomenon and what it does to humans. Why has it inspired so many artists to make paintings and write novels about it? Not only do many musicians sing about the blue hour but there is also a music genre originating in the Southern part of the United States which expresses the sorrows and sadness of African Americans. Blue as a color is used as a metaphor to refer to specific emotions, a particular type of mood produced by certain circumstances and experiences. When we feel sad, we are "feeling blue." When we have the "Monday morning blues" we feel tired when we have to get up early and back to work. Colors, it seems, apply to objects and substances (such as blue cars and blue sky) as well as concepts and ideas. They are also linguistic ways to convey messages and meanings. Even if they are not really there in the objective world, they help us to imagine and interpret our world in particular ways.

In this book the role and significance of color will be examined in two areas of human activity: healthcare and bioethics. While colors surround us, are directly experienced, and often enrich our perception of the world, they do not receive much attention in medical and care activities, and even less in ethical analyses. Usually, color is regarded as secondary and trivial—a subjective impression deemed less important than objective observations and findings. Nonetheless, as discussed in this book, color has played and continues to play an important role in healthcare, not only in diagnostic but also therapeutic endeavors. The same is true for ethics. Historically, ethics was clearly demarcated from aesthetics, with color relegated to the domain of emotions, feelings, intuitions and subjective experiences, while ethics was characterized by rational arguments and deliberation. In present-day bioethical debate, this distinction can no longer be upheld since, in practice, colors (particularly black and white) carry moral connotations and interpretations that influence ethical judgments before they are rationally articulated.

1.2 Color and Medicine

While contemporary medicine is regarded as an objective and scientific enterprise, color plays a special role in healthcare activities. For a long time, diagnostic means were limited and doctors relied on inspection and observation to clarify the ailments and complaints of patients. When the face is extremely pale and dusky, it may together with other symptoms indicate imminent death, described by Hippocrates as the *facies Hippocratica*. A bluish-purple coloration of the skin on the other hand is, according to Hippocrates, indicative of respiratory problems.

Even today, medical students are taught to take a medical history and perform a physical examination. This involves first of all a systematic inspection of the body of the patient, and its various parts. Colors of the body such as redness, cyanosis (blue), jaundice (yellow) and pallor may give clues for possible diagnoses. Excretions may have various colors, indicative for specific problems. Brown or orange urine may suggest liver disease, while red urine can be caused by certain foods but may also be a sign of a serious health problem (loss of blood due to kidney stones, infection or cancer). Stool may a have range of colors but black and red require medical attention. When a person's teeth become red-brown it indicates porphyria, a very rare genetic liver disease.

Since Ancient times, medical theory was based on the primacy of colored substances for health and disease: the four humors, namely blood, black bile, yellow bile and phlegm.

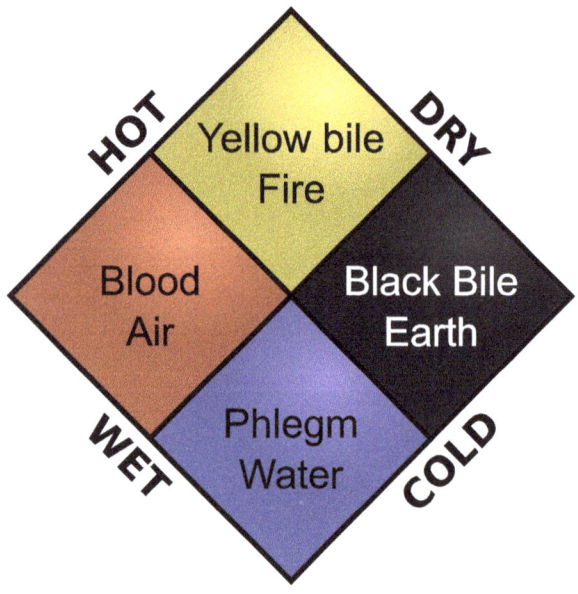

Fig. 1.2 The four humors, Tom Lemmens (2013). Wikimedia, https://commons.wikimedia.org/wiki/File:Humorism.svg#/media/File:Humorism.svg, CC0 1.0.

A healthy constitution depends on the balance of these body fluids, which also determine a person's predisposition toward disease. If the balance between them is disturbed, and one of the humors is excessive or deficient, specific diseases will develop. If black bile dominates, melancholy will be the result; if there is an excess of yellow bile, jaundice or icterus will be observed. The humoral doctrine which determined medical thinking and acting for centuries was not an isolated theory but connected to a wider cosmology, relating the basic constituents of humans to fundamental elements of the universe: air (blood), fire (yellow bile), earth (black bile) and water (phlegm). The basic fluids, furthermore, were assumed to have psychological impact since they determined different temperaments: sanguine, choleric, melancholic and phlegmatic. This ancient theoretical framework assumed a close connection between colors and the physical and psychological constitution of humans. This is noticeable for example in the naming of diseases. It also explains the linkage between pigments and medication. In the past, many pigments were used for painting and coloring objects but also as drugs for a variety of ailments. The yellow pigment gamboge, for example, first imported from South-East Asia, became in the seventeenth century a popular medicine against rheumatism, scurvy and other illnesses (St Clair 2016). The search for new synthetic dyes in the nineteenth century was a major catalyst for the emergence of the pharmaceutical industry.

In medical history some diseases are easily recognizable because of typical coloring: scarlet fever, rubella (from the Latin *ruber*: red), erysipelas (from the Greek *erythros*: red), yellow fever, albinism (from the Latin *albus*: white), blue baby syndrome, ochronosis (from the Greek *ochros*: pale yellow), leukemia (from the Greek *leukos*: white), melanoma (from the Greek *melas*: black), and glaucoma (from the Greek *glaukos*: gray, bluish-green) Cholera was known as the 'blue death' because dehydration turns the skin bluish. One of the most severe medical disasters of the past, the fourteenth-century plague, became known later as the Black Death, presumably because the infection caused black boils in the armpits, neck and groin due to internal bleeding. The plague pandemic of 1348–1353 in Western Europe was initially called the Great Pestilence, and in Latin *atra mors*, dismal or terrible death. Because *ater* also means black or dark, the expression was later mistranslated as black death (Benedictow 2004). In this interpretation the color term does not refer to the hue that is visible in patients but first of all invokes a symbolic meaning: death, earth, darkness, grief and hell. The Latin word *ater* is associated with black bile, a worrying and matt black, in contradistinction to *niger* which indicates shiny, glossy black and has positive connotations of respectability, austerity and authority (Pastoureau and Simonnet 2005).

The 'green disease' is identified in the sixteenth century as a specific ailment of young teenage girls, and therefore also called the virgin's disease (Starobinski 1981). The subjects have a host of symptoms, from fatigue, lack of appetite to palpitations, paleness and absence of menstruation. The explanation is that the symptoms are caused by excess of blood since it cannot be released through menstruation. The remedy for the disease therefore is sexual activity so that the proposed cure is marriage, followed by pregnancy and children. The disease is also known as chlorosis (from the Greek term *chloros*, pale green). It is likely that the girls are suffering from anemia; they are in fact pale rather than green. The color evokes the traditional symbolic meanings of green: youth and love but also hope, destiny and fertility (Pastoureau 2014).

These examples show that associating colors and diseases has different functions. First of all, the color may be a sign that something is wrong with the person. The color of the body or body parts may reveal that normal functioning is disturbed or that pathology within the body is manifested at its surface. Blue skin for example is an indication that oxygen levels in the blood are low. At the same time, color conveys a specific meaning and is used as a symbol. It may or may not refer to visible hues but, above all, evokes emotional responses. Death is black because it means extinction and darkness whether or not patients have a specific coloration of parts of their bodies. A particular combination of symptoms is labeled as green disease, not because sufferers become green but because they are young, female and just entering reproductive age. Colors furthermore have a third function which is also particularly relevant in healthcare: they can act as signals used to communicate certain ideas and influence practices. Hospitals use various color codes. Code 'blue' means that there is a medical urgency; a patient has a cardiac arrest and immediate resuscitation is required. Code 'red' indicates that there is a fire or smoke in the healthcare facility. Code 'black' became especially relevant during the Covid-19 pandemic. It is declared when hospitals are at capacity; there are not enough beds to treat every patient who needs it. This has serious ethical implications since physicians have to choose who will be prioritized for treatment. Color codes were also used by policy-makers to indicate the threat level of the coronavirus. Red refers to the most dangerous level, and green indicates safety. However, countries implemented various colors for Covid-19 alerts. Most countries used a four-color system with amber (caution) and orange in between red and green. Some countries revised their color-coded system: the United Kingdom, for instance, moved from a five-level to a three-level system (Shendruk 2021). The European Union adopted a common color-code to coordinate travel restrictions with red, orange and green, and published a weekly color map of countries. The coronavirus dashboard launched by the World Health Organisation used red for the number of deaths, and shades of blue for the number of confirmed cases; the darker the blue, the more cases.

1.3 Color and Ethics

Since the Covid-19 pandemic, it has been argued that the color of coronavirus disease is black. This argument illustrates that color is not only used as a sign, symbol and signal but also as a means of expressing moral values. Covid-19 is associated with the color black because the disease disproportionately affects older Indigenous, Latino and Black Americans: among these populations, rates of infection, hospitalization and death due to the coronavirus are significantly higher than among White populations, with death rates approximately two to three times greater. Also, the social impact of public health measures on these racial and ethnic groups is stronger: these populations face higher rates of unemployment, reduced ability to work remotely, and are more likely to be frontline workers, who are at greater risk of exposure. Racial and ethnic populations in the United States also have higher risks of severe illness and death from the coronavirus due to pre-existing health inequalities rooted in longstanding structural racism and discrimination (Garcia et al. 2021). Describing the color of Covid-19 as black underscores that not everybody is affected by the virus in the same manner and with the same severity. The pandemic exposes and intensifies the existing inequalities in health and society; Black and Brown populations are more vulnerable because of systemic disadvantages, including lack of access to healthcare, unsafe living conditions, limited employment opportunities and environmental degradation. Labeling the disease as black serves as an outcry against structural racism, and a moral call to action to address these disparities. Ironically, the coronavirus itself has no color. Although its images are often presented in red, the virus is too small to interact with visible wavelengths of light, and therefore lacks color entirely (Siegel 2020). By presenting it in red, the meaning of this color as danger and harm is invoked.

The association of color with morality is demonstrated in multiple studies. White and black, especially, are regarded as symbols of moral purity and pollution. Black is associated with ideas of dirtiness, impurity and immorality. It is the color of the night, darkness, uncertainty and potential danger; it can also contaminate other objects. Moral connotations are also attached to other colors. Blue, which is now the preferred color in Western cultures, was for a long time not appreciated. In Ancient Rome, it is the color of barbarians, and in Ancient Greek texts blue is absent. In the Middle Ages it became a divine color, widely used in stained glass in churches and adopted by the kings of France. During the Reformation it became one of the "worthy" colors, with moral discourse promoting black, gray and blue. Colorful objects and clothing were regarded as extravagant, while blue was associated with calmness, modesty and serenity. Blue, along with black and white, was considered a more worthy and virtuous color. In addition to its historical associations, blue holds significance as a color of consensus: today, it is used in the emblems of the United Nations and the European Union (Pastoureau and Simonnet 2005).

The idea that colors have a moral significance has a long tradition. In Ancient Rome a distinction is made between good and bad colors. The first group (*colores austeri*: white, red, yellow and black) are honest colors because they are dignified, decent and restrained. The other group (*colores floridi*: vivid colors) are frivolous, false, vulgar, merely decorative and thus dishonest. According to Plinius and Seneca, these last colors are extravagant and decadent; they are usually imported from abroad, and have an exotic and oriental origin. Instead, the Romans favored simplicity and austerity, and thus traditional colors (Gage 2013). Since the Reformation, the belief that some colors are more worthy than others has remained strong. Bright colors, which attract the eye and capture attention, divert from what is virtuous (Pastoureau 2010). Clothing should be somber, simple and plain so that it expresses humility, sincerity and austerity, with priority given to black, gray and blue. Protestant chromoclasm, the effort to expel vivid colors from public life, is visible in the paintings of Rembrandt: the physicians in *The Anatomy Lesson of Dr. Nicolaes Tulp* (1632) and the officials in *Syndics of the Drapers' Guild* (1662) are all dressed in black, like almost all militiamen in *The Night Watch* (1642). Some have argued that Western culture, more than other cultures, is characterized by chromophobia, leading to recurrent trends of marginalizing and devaluing color, diminishing its significance, and denying its complexity (Batchelor 2000). First of all, color is regarded as dangerous and pathological; it is seductive and can corrupt and misguide the mind because it is not directed at the intellect but at the senses. However, it cannot be avoided or ignored since the world is colorful, but rather must be contained and subordinated in order to control the emotions that it incites. Chromophobia also exists because bright colors are not refined and sophisticated; they are usually exotic and imported. Johann Wolfgang von Goethe argued that "savage" nations, uneducated people and children have a predilection for vivid colors (Batchelor 2000). Colors are therefore often defined by "otherness": they are feminine, oriental, primitive, infantile and vulgar. This is related to the Western experience of colonization: "Color in the West became attached to colored people or their equivalents…" (Taussig 2009, 16). Most desired colors, and especially pigments, came from exotic places beyond Europe: for example, the emergence of blue in the eighteenth century as a popular color is dependent on indigo plantations in Central American and Indian colonies, and thus interconnected with slavery.

Another reason for chromophobia is that color is often regarded, especially in philosophy, as trivial. Immanuel Kant, to mention one example, argues that in painting, sculpture, architecture, and in all fine arts, the drawing, the design is essential. It is the form that pleases and that gratifies the sensation of beauty. Colors merely illuminate the outline and make it livelier; they may be charming but do not make a piece of art beautiful (Kant 2001, §14). In many philosophical treatises color is interpreted as a secondary quality of experience. They argue that the word "color" is derived from the Latin verb *celare*, meaning to hide, cover or conceal. Color is like a second skin, make-up, an envelope that covers beings and things, a surface rather than the substance itself. Its trivial nature is accentuated in the enduring debate of design versus color in art (Riley 1995). Design or form is rational, structured, honest, reliable and an example of moral rectitude, while color is identified with the emotional, rhapsodic, formless and even deceitful. Composition, line, subject or perspective are more important than color because color is not the creation of the human mind or the expression of an idea. For the architect Le Corbusier color is intoxication; it is suited to simple "races, peasants and savages." He prefers white because it is no color in his view and thus represents order, purity and truth: "White is clean, clear, healthy, moral, rational, masterful…" (Batchelor 2000, 46). The distinction between color and shape is also applied practically in the Rorschach test—the psychological test of someone's personality using abstract inkblots. A powerful response to color indicates emotional instability and impulsiveness, whereas a strong response to shape suggests control or balance between emotional and intellectual life (Riley 1995).

1.4 The Outline of the Book

Before the relevance and significance of colors within healthcare and bioethics can be examined, it is necessary to explore various ideas and theories about the nature of color. Since colors are everywhere in our surrounding world, they are immediately experienced by all human beings, except those with particular forms of colorblindness. Not only do objects and entities in the experienced world possess the property of color, but we ourselves are colored. Our bodies have different colors that can change over time and across different environments. Moreover, colors can reflect varying states of health and indicate certain diseases. We use color terms to refer to different states of mind and varying moods, to make social distinctions, and to morally judge behaviors and practices. The omnipresence of colors in everyday life has stimulated reflection since the dawn of philosophy. Following the Introduction, the second chapter will discuss ideas about what color precisely is.

1.5 The Nature of Color

Reflections on the nature of color have often oscillated between objective and subjective interpretations. Especially since the experiments of Isaac Newton in the 1660s, colors have been regarded as objective realities. Newton relates color to light, following Aristotle's suggestion that light is the activator of color (Gage 1999). When visible light is refracted through a prism, it contains seven fundamental or spectral colors (red, orange, yellow, green, blue, indigo and violet). Each color has a specific length and frequency of electromagnetic waves. Red has the longest wavelength and violet the shortest. The surfaces of objects in the world absorb and reflect these waves so that specific colors become visible. The chemical constitution of materials determines what wavelengths are absorbed and reflected. If all wavelengths are reflected, we see white; if they are all absorbed, we see black. If the longest wavelengths are reflected, and the others absorbed, we see red. The studies of Newton have promoted physicalistic theories. Colors are physical properties of material bodies and entities, and can be measured since they are written in the language of mathematics (Romano 2020, 66). Another scientific theory of color is based on neurophysiology. It emphasizes perception: wavelengths of visible light are only colors when we see them. Colors are produced and constructed in the visual system and do not exist outside of perception. The retina has two types of photosensitive cells: rods (to distinguish between dark and light) and cones (to distinguish among colors). There are three types of cones, sensitive to short, medium or long wavelengths, and usually labeled as blue, green or red. Light waves activate the rods and cones which then send messages to the visual cortex of the brain, that interprets the physical sensation as the perception of colors, depending on what kind of cones are activated. Without cones wavelengths have no color. Neurophysiological theories thus provide an explanation of color in anatomical and physiological terms but they are subjective theories in the sense that color is not a property of objects in the outside world but completely produced within the subject's perception.

One conclusion so far is that objective as well as subjective interpretations have the same effect: they assign a particular location to color; it is either outside in the physical world or inside in our brains (Romano 2020, 135). They reduce colors to either sensations and sense data that stimulate the human visual system or perceptions that are constructed in the visual system. Milk itself is not white; because its molecular composition reflects all wavelengths of visible light, our visual experience of milk is white. Since the wavelengths reflected by the milk activate the neuronal system, we have the perception of whiteness. In both cases the color does not exist in the outside world unless we take the physicalist view that identifies colors with wavelengths of light. Usually, colors are regarded as sensations or perceptions rather than as experiences which relate objective and subjective elements. This last view is elaborated in the theory of phenomenological realism which regards color as a relational property. It is not an intrinsic property of an object (determined by its physical characteristics) nor an idea or mental construct (determined at neuronal level), but a relational property that brings together the object and its environment as well as the perceivers. Colors have a reality in the phenomenal world which is partly independent of human perceivers, and it is also more than a private mental state in the perceivers. This relational theory is influenced by an ecological view of colors that attributes specific functions to perception. The aim of perception is to detect certain characteristics of the environment that are useful for the survival of a species, and to discriminate between beneficial and harmful objects.

Against this phenomenological backdrop, perception is not simply detection and recording of objects in the world but it structures and patterns the world into visual fields. The world in which it takes place and which it reveals is not the objective, physical or biological world but the world of immediate, lived experience. In the immediacy of the experience there is no distance between humans and world; being human is fused with the world. We exist, live and act in the life-world (*le monde vécu*) before we can make this world the object of scientific reflection and analysis. The cognitive relationship between knowing subject and known object is preceded by the intertwining of human beings and world, and is first of all a perceptive relationship. Perception takes place at a pre-reflective level; it brings us into contact with the world that is prior to scientific knowledge; according to Maurice Merleau-Ponty, perception, in contrast to knowing, is a living communication with the world that makes it present to us as the familiar place of our life (Merleau-Ponty 1945, 64–65). In his view, a color is felt and the body is responding before we are even aware that we see it. Colors have "*signification motrice*": they are touching and moving us (Merleau-Ponty 1945, 243 ff).

1.6 The Power of Color

The third chapter explores the suggestion that colors have particular effects on human beings. Since color is a multidimensional phenomenon, it is studied from a range of perspectives: physical, physiological, neurological, medical, psychological, social, cultural, linguistic and aesthetic ones. Color is not simply a surface or ornament but it can transmit and communicate codes, prejudices and normative judgments. It influences language, behavior, imagination and sentiment (Pastoreau and Simonnet 2005). These effects of color are particularly expressed by artists. For Henri Matisse, for example, colors are powers; he uses colors in his paintings not to transcribe nature but to express emotions. According to Wassily Kandinsky, color is a power which directly influences the soul (Riley 1995, 136, 142). Color is a language without words and it can directly address our emotions and feelings (Street 2018).

That colors have an impact on emotions and feelings is shown in numerous studies. College students in the Unites States, for example, associate red with excitement, orange with distress, yellow with cheerfulness, and blue with comfort and security. Colors can evoke positive as well as negative emotional responses. Green has a retiring and relaxing effect, and gives the impression of refreshment, naturalness and quietness but it is also associated with tiredness and guilt (Kaya and Epps 2004). Cross-cultural studies describe similar patterns in a variety of countries. Blue is the most highly esteemed color; least preferred are grey and black. The strongest responses are associated with red: it is heavy or intense in feeling, and generally regarded as an active color (Adams and Osgood 1973). If colors are connected with meanings and emotions, this may have implications for psychological well-being and functioning. The color red is positively related to failure, and negatively to success, while green is positively related to success. These connotations are also expressed in ordinary language with negative references for red ("in the red," "red tape," "red herring") and positive ones for green ("green light," "green fingers," and "greenback") (Moller, Elliott and Maier 2009).

Human performance is the subject of a range of studies in various settings. People make more proofreading errors in interior offices which are white compared to blue and red offices, even if they mostly prefer to work in white and beige offices which are often regarded as the least distracting colors (Kwallek et al. 1996). The learning performance of students is best when the walls in their study room are blue. Red-colored walls have a negative impact on intellectual activity by impairing concentration (Al-Awash et al. 2016). The negative effect of red is also demonstrated in IQ test performance. In contexts in which an achievement is expected, perception of red impairs performance, particularly when cognitive analysis, mental manipulations and flexible processes are required (Maier, Elliott and Lichtenfeld 2008). On the other hand, blue enhances performance of a creative task. If creativity and imagination are required (for example, in the development of a new product or a brainstorming session) blue is more beneficial than red (Mehta and Zhu 2009). Effects of the color red and its associated meanings seem to depend on the context. When this is competitive, red positively influences the outcome of a contest. When the context is relational, red enhances attractiveness.

These studies of the effect of colors usually have practical implications. An illustrious example is the finding that waitresses in restaurants wearing red receive more frequent and higher tips from male (not female) customers than those wearing other colors (Guéguen and Jacob 2014). This finding indicates how waitresses may increase their income. The assumed effects of colors on human emotions and behaviors is especially examined in the marketing industry. Giving brands of products a particular color not only helps to recognize the brand but also to establish a visual identity which communicates a certain image; it creates a distinctive personality of the brand. Red associates a brand with excitement, white with sincerity, and blue with competence (Labrecque and Milne 2012). Decisions to purchase a product are not only based on brand, price and quality but also on color. Consumers ascribe a particular meaning to the color of products. The effects of colors are furthermore studied in relation to food. We immediately appraise the quality of meat, fish and fruit by their colors. Color is often the most powerful visual aspect of food packaging, intended to influence consumer decisions. Red packaging is associated with hot flavors, and green with nature and environmental friendliness (Yu et al. 2021). Another question is whether the color of food influences taste and the perception of flavor. Studies document that red food coloring has an effect on the perception of sweetness. A finding that drew substantial popular attention was that people eat less when food is served on red plates, possibly because the color red signals danger and prohibition, consequently inducing avoidance behavior (Bruno et al. 2013).

1.7 Color and Healthcare

The role of color in the context of healthcare is the subject of Chapter 4. As discussed earlier, colors are used in medicine to classify diseases and diagnose disorders. Specific color tests have been developed for diagnostic purposes. Examples are the litmus test to determine the acidity of a substance, and the gram stain to identify microorganisms. But colors also play a role in the fields of medication, therapy and care environment.

Substantial efforts have been invested in examining the effect of the color of medication on its use, popularity and efficacy. The response to medication is not only determined by its chemical composition but also by other factors. How it looks like (i.e. its preparation form, size and color) generates certain expectations concerning action and strength. Black and red capsules are perceived as more potent than orange, yellow, green and blue ones. White capsules are generally perceived as weak (Sallis and Buckalew 1984). A review of the literature concerning the color of drugs with stimulant and depressant effects concludes that red, yellow and orange are associated with a stimulant effect, blue and green with a tranquillizing effect (De Craen, Roos, De Vries and Kleijnen 1996). These associations are generally consistent across countries (Amawi and Murdoch 2022). Traditionally, medication was not colored but in the 1970s the new technology of producing soft-gel capsules made colorful drugs possible. Nowadays, capsules and tablets can have thousands of color combinations. Usually, the coloring of medication is defended with several arguments: it helps consumers to recognize medication, and for pharmaceutical companies, coloring frequently plays a role in marketing. Another argument refers to the supposed power of colors. If colors have particular meanings and raise specific expectations about their efficacy, then responsivity of patients to medication will be better when its color corresponds with its intended effects (i.e. use red for speedy relief, and blue for sleep and calmness).

While colors influence the working of medication, they can also impact patients who do not use drugs. Studies indicate that colors can affect the perception of pain. Red has a stimulating effect and can intensify pain, compared to green and blue. White is a pain-reducing color (Wiercioch-Kuzianik and Babel 2019), associated with purity, cleanliness and hygiene, and has sedative properties. The idea that colors may act as therapeutics is clearly manifested in the history of medicine, particularly in approaches to smallpox. In many cultures, red was used to offer protection against this disease (Hopkins 2002).

A belief in the healing powers of colors has promoted chromotherapy. When colors are a physical phenomenon, it can be supposed that each color has a specific wavelength, and thus vibration, which affects the body and specifically its chemical constitution. Diseases can therefore be healed by color vibrations, and for each disease a specific color can be used (Klotsche 1992). Knowing, for example, that red is stimulating, it activates digestion and the liver; blue, which is soothing, has a catabolic effect and reinforces the immune system. Advocates of chromotherapy have developed theories to specify the action of colors upon different organs and systems of the body. They argue that the advantage of chromotherapy is that it leaves no harmful residues in the body, in contrast to medication (Anderson 1990). Another therapeutic use of colors focuses on light, applying artificial light for a variety of conditions, an example being the exposure of newborns with severe jaundice to blue light (Stokowski 2011).

Another use of colors in healthcare is within the environment of patients. The rationale is that the interior design of hospitals and other healthcare facilities should contribute to the recovery process of patients and to enhance the well-being of all users of these facilities. Colors may contribute to the positive experience of these surroundings. They are, first of all, important for navigation and spatial orientation. But the assumption that colors have physiological and psychological effects has more fundamental implications. Exposure to green colors, for example, is associated with improved feelings of well-being. The most common color in hospitals used to be 'hospital green' or 'spinach green.' (Pantalony 2009). The dominance of green in hospital settings is increasingly criticized, and since the 1970s many other hues are used (Olgunturk et al. 2021).

1.8 Color and Bioethics

Chapter 5 will elaborate the connections between colors and bioethics. Colors have often been associated with normative judgments. In medieval moral theology, the seven deadly sins were each represented with a color, leading to the idea that ethics should avoid association with any of them. Pastoureau (2019) argues that the Protestant Reformation introduced a moralistic approach to colors in public life, distinguishing "worthy" from "unworthy" colors. The first group (white, black, brown, grey and blue) are the expression of certain values such as soberness, discreteness and dignity. The second group, including yellow and green, were deemed disgraceful and improper, and almost disappeared from public life in some parts of Europe. However, the idea that some colors, especially bright ones, are transgressive and morally inappropriate is much older, and already expressed in classical Antiquity. They are regarded as misleading since they attract the eye and capture attention, directing our mind to the surface of things rather than their essence. Distrust of colors is related to the idea that the human being is distinguished from other living beings since it is a rational animal, characterized by discursive thinking, explanation and argument. Color is relegated to the domain of emotions and subjective impressions. In this view, the vibrant hues of the surrounding world obscure a more fundamental reality which can only be discovered and analyzed by the mind. Therefore, colors are subtle deception; they are merely external, ornamental and decorative. This value judgment about colors was regularly connected to another one: that colors are extravagant and decadent. Using many colors is not a matter of refined taste and civilization. It indicates that the moral values of a society are declining, and that traditional values such as simplicity and honesty are no longer cherished. This association is partly explained by the historical reliance on exotic and costly pigments, which were imported from abroad, reinforcing the idea of color as foreign and indulgent.

That colors have a moral value is clear in the hierarchy which many societies apply to them. Batchelor (2000) argues that cultures often oppose colors with white, regarded as colorless. White is associated with innocence and purity (Pastoureau and Simonnet 2005). It is a guarantee of cleanliness and hygiene. The moral value of colors is furthermore evident in their use to articulate social divisions and distinctions. In the past, numerous societies had color codes and stringent regulations for the application of color in public life (Pastoureau 2017). Social classes are indicated by the colors that they are allowed to use for their clothing. But this moral value of colors has become problematic when it is applied to people themselves. While in the past, different groups of humans have often been identified by the color of their skin, the classification of the German physician and anthropologist Johann Blumenbach in the 1770s became highly influential. He distinguished five varieties of the human species ("races") according to skin color: Ethiopian (black), Caucasian (white), Mongolian (yellow), Malaysian (brown) and Amerindian (red). Colors were understood to vary due to geographical factors, such as climate. Blumenbach argued that the differences between these varieties are so small and gradual that it is almost impossible to make sharp distinctions. At the same time, he strongly opposed any hierarchy among the varieties, rejecting the suggestion that some are superior and others inferior (Pastoureau 2019). Nonetheless, this is exactly how his ideas were interpreted and elaborated in theories of scientific racism. In classifications of people, skin color was associated with character and moral worth. Particularly, black was connected to evil and negativity, while white was believed to be superior (Jablonski 2012).

The pervasiveness of moral associations of white and black has now become a major topic of concern in ethical debates. The Covid-19 pandemic—which disproportionately affected populations of people of color—and the Black Lives Matter movement have placed racism, structural injustice, discrimination and vulnerability more center stage in bioethics (Russell 2022). It is recognized that racism is a barrier to health and healthcare of non-White people. Since it is an imperative in healthcare ethics to prevent harm to patients, racism and the concept of race should be the focus of bioethics. That requires an analysis of the contextual and structural dimensions of health and diseases, and also, as Russell (2022) points out, an awareness that all subjects of bioethical inquiry are racialized. She suggests that bioethics itself, as it has emerged as a new discipline since the 1970s, is based on an underlying principle of White supremacy, i.e. the idea that white lives are of greater value than those of people of color. The theoretical framework of bioethics, with its focus on autonomy, consent, transparency and risk assessment, presupposes individual citizens who are independent and free to make decisions. However, this framework largely ignores non-White people who are disadvantaged and vulnerable because of social, economic and environmental conditions. In these analyses, whiteness is often rendered invisible as a racial category, and White people are seen as a neutral social group. As a result, race is primarily applied to non-White people, while whiteness becomes the normative standard from which deviations are assessed. White is then equated with being human, and the embodiment of universality (Dyer 2017). This is reflected in the practice of "race norming", i.e. the adjustment of test scores to account for the race of people who are tested. The assumption is that the physiology of white bodies is the norm and that outcomes for people of color need to be corrected because their physiological capacities are considered inferior (Braun 2014). While race norming was prohibited by law in the United States in 1991, similar criticisms are now directed at the use of algorithms in healthcare, which often disadvantage communities of color (Ledford 2019).

Current debates about race and racism are shifting the focus of attention from black to white. Whiteness itself has become problematic with criticisms of White superiority and White privilege. Movements as 'wokeism' are motivated by resistance to the power of White men (Weyns 2023). The great replacement theory, popular among conspiracy thinkers, regards White people as an endangered species; White populations are systematically replaced through mass immigration of people of color, and intermingling between Whites and people of color (Rose 2022). White has become a metaphor for a world that is disappearing while for anti-racists it is a symbol of power and privilege. The obsession with whiteness, however, shares the same prejudices as the pejorative connotations of blackness, attributing moral qualities of superiority and specialness to a specific color. They reflect anxieties and fears about a world which is changing due to demography, immigration, wars, climate, disparities and structural violence. Both keep alive the ideology of colorism: discriminatory treatment of individuals based on skin color. They forget that colors have power to condition our behavior and way of thinking, but that the colors themselves are ambiguous and can induce various associations. For example, in some cultures white is the color of death and mourning while in Western culture black is also the color of elegance and seriousness. Furthermore, colors present themselves in a range with varying hues and intensities. White people are not really white, unless they are ill (anemia and tuberculosis). The same goes for dark colors which present themselves also in a huge diversity. This is directly visible in the project of Angélica Dass, who made 4,000 photographic portraits in order to document the uniqueness and diversity of the color tone of faces (Dass 2023).

Fig. 1.3 Angelica Dass Retratos, *Proyecto Humanae Valencia*. Color labels like black, white, yellow or red are inadequate to capture this diversity. Photo by LOLAOMI (2014), Wikimedia, https://commons.wikimedia.org/wiki/File:Ang%C3%A9lica_Dass_retratos.jpg#/media/File:Ang%C3%A9lica_Dass_retratos.jpg, CC0 1.0.

Additionally, the moral use of the contrasting colors white and black is criticized from an evolutionary point of view. Skin colors vary because different amounts and kinds of the pigment melanin are produced to protect skin against damaging solar radiation. Initially, all humans were dark skinned. When *homo sapiens* moved out of sub-Saharan Africa at least 60,000 years ago, gradually their skin became lighter in the process of adaptation to life in the northern regions of the globe (Jablonski 2012). DNA studies show that the hunter-gatherers in Western Europe had a dark skin that only slowly lightened in order to facilitate the production of vitamin D (Posth et al. 2023). The moral problem lies not in the color itself, but in the associations it evokes—especially the tendency to rank people and attribute specific character and moral worth based on color. This is further evident in historical examples, such as the way the British once viewed the Irish as black (Dyer 2017). Also, in US immigration policy, stringent laws were first drafted to counter the influx of Chinese and Japanese immigrants, and later to restrict immigrants from Central and Southern Europe. These laws were fueled by fears that non-White immigrants were going to replace White Americans (Jones 2021).

1.9 A Colorful Bioethics

Recognition that colors are associated with moral appreciations and that these associations need critical analysis has implications for the subjects addressed in bioethics. This recognition not only requires that certain topics such as racism, structural violence and discrimination should be higher on the agenda of contemporary bioethics, but also demands the expansion of the field of ethical inquiry. The relevance of color demands that bioethicists are aware of racism and colorism as determinants of well-being and disease but also, as the example of Covid-19 demonstrates, that they focus attention on the underlying mechanisms that disadvantage people of color. It is therefore important not only to expand the agenda of bioethics with the addition of more relevant topics but also to broaden its approach. Ethical examination should be reorientated towards contextual and structural conditions rather than focus on the individual perspective of rational and autonomous persons. This means that a broader framework of ethical approaches and principles must be employed than is currently applied. Commonly used ethical principles such as respect for individual autonomy and non-maleficence are engaged as universally applicable, assuming that they are equally appropriate for all human beings regardless of their situation and predicament. Contemporary bioethics perceives itself as colorblind. It assumes that when colors are not 'seen' or simply regarded as irrelevant or trivial particularities, differences in reality, and especially differences among people, no longer exist. Erasing color as a relevant ethical consideration removes the possibility of exploring why disadvantages and injustices prevail, and analyzing why people are affected and treated differently (Mesman 2021). Acknowledging that color is a relevant consideration in health and disease, and that its power necessitates critical analysis that goes beyond the usual ethical principles of respect for individual autonomy and non-maleficence, results in a conception of bioethics that is genuinely global, i.e. relevant for all peoples, ethnicities and cultures around the world.

Furthermore, consideration of color and its relevancy in healthcare ethics focuses attention on the relationship between ethics and aesthetics. Though traditionally connected in Western philosophy, they are nowadays mostly separated. Ethics is concerned with what is good and right; it aims to determine what ought to be done, and it uses general principles to guide rational arguments and deliberations. Aesthetics is concerned with beauty; it involves the senses, particularly seeing, when colors are concerned. Because the senses are considered as less reliable than reason, aesthetics is regarded as a matter of affection and intuition, thus personal taste. The term 'aesthetics' is derived from the Greek *aisthánomai* which means perceiving, feeling and sensing. Aesthetic judgments are based on human sensitivity, imagination and intuition and as such assumed to be sources of error. It seems that the traditional distinction between the profound and the superficial is at work here: ethics is the search of goodness, proceeding from rational arguments and deliberation, and focused on identifying reasons for and against acts and decisions. In contrast, aesthetics is driven by emotion and intuition, focusing on subjective experiences of what appears to be attractive or beautiful.

The common view of bioethics as an abstract system of moral principles and rules, working on the basis of arguments and rational reflection and with clear procedures for decision-making is nowadays increasingly criticized. It is argued that principles require continuous interpretation and cannot directly be applied to moral dilemmas in order to provide clear-cut answers. It is also argued that ethical decision-making takes place within concrete contexts and practices and is therefore not abstract but drawing on the moral experiences of the persons involved. Moreover, moral judgments and decisions are not merely rational but influenced by values and emotions which determine what is relevant and significant. These criticisms have articulated the crucial role of moral perception in ethical discourse. Before a moral judgment can be delivered and before moral reasoning and rational deliberation can take place, particular situations must be perceived as morally significant. Such perception requires moral sensitivity and experience but is also facilitated by the imagination that expands our perspective and situates ourselves in the circumstances of other people. Rehabilitating the role of perception in ethics re-establishes its connection to aesthetics as the science of sensory perception (Macneill 2017). This connection is furthermore reinforced with the new appreciation of emotions and feelings in moral reasoning and deliberation. Though usually discarded and considered as obstacles to rational decision-making, cognitive psychologists nowadays argue that most moral judgments are made through an intuitive process based on emotions and feelings, which operates more quickly than reasoning. A conscious reasoning process, such as that used in moral analysis and deliberation, is employed after a moral judgment is made to justify this judgment (Haidt 2001).

When it is concluded that ethical reflection and moral deliberation are not entirely rationalistic processes but connected to intuitions and emotions, the relevance of color for ethics must be reconsidered. It is not a trivial side issue in our dealings with the surrounding world but it presents this world in specific ways and is omnipresent in the interactions and communications between people. At the same time, it conveys particular emotions, values and judgments, and therefore influences the intuitive process of making moral judgments. Perceiving a specific color or range of colors produces an immediate and intuitive experience which generates a value judgment prior to rational deliberation. In this way, ethics already starts in the concrete experience of perceiving which then necessitates critical examination and explanation with the help of systematic theory and moral reflection. Perhaps this is what Emmanuel Levinas has in mind when he writes that ethics is an optics, a way of seeing (Levinas 1961, 8, 15).

1.10 Conclusion

The world around us has an infinite and dynamic variety of colors which make our visual experiences beautiful, enjoyable and wonderful. Colors have an effect upon us; they can be stimulating or irritating, they can attract or make us nervous, bring us to rest or encourage reflection or meditation. Experiencing colors not only has aesthetic or psychological effects but also influences normative valuations since they are associated with positive and negative values and intuitive judgments. Because colors are not inert or indifferent but have the power to touch or move us, it is important to reflect upon their role in ethical debates of health and disease, life and death. This book will examine color in the context of healthcare and bioethics. It will argue that color fundamentally is an experience rather than a sensation or perception. It is not an illusion constructed in our mind, nor a physiological phenomenon in the outside world but the experience of relationships between perceiving subjects and their life-world. This experience is active since it makes the world meaningful and relevant, shaping and structuring it in order to allow us to orientate ourselves, to make us feel at home, to fulfill our intuitions, and to flourish. Interpreting color as a relational phenomenon explains what is often referred to as the power of color.

This power is noticeable in numerous areas of human activity, though perhaps least of all in healthcare and bioethics. As argued in this book, there is a fear of color especially in Western culture that clarifies why color is often considered as trivial, superficial and irrelevant. Nonetheless, this book argues that color plays a significant, and mostly positive role in healthcare practices, particularly in diagnostics and therapy. In bioethical debates, little attention until recently has been given to the relevance of color. In the past, ethical theories articulated the color of the skin as indicator of differences between human races. Because of this racial and racist history, color in bioethics has negative connotations and the prevailing ideology is that color should not play a role in healthcare interactions. The argument in this book is that color in bioethics cannot be denied or ignored. On the contrary, it should be acknowledged as a positive experience that connects our appreciation of both the goodness and beauty of the surrounding world. But first, it is essential to ask: What is color, fundamentally?

References

Adams, F. M., and Osgood, C. E. 1973. A cross-cultural study of the affective meaning of color. *Journal of Cross-Cultural Psychology* 4 (2): 135–156, https://doi.org/10.1177/002202217300400201

Al-Ayash, A., Kane, R. T., Smith, D., and Green-Armytage, P. 2016. The influence of color on student emotion, heart rate, and performance in learning environments. *Color Research and Application* 41: 196–205, https://doi.org/10.1002/col.21949

Amawi, R. M., and Murdoch, M. J. 2022. Understanding color associations and their effects on expectations of drug efficacies. *Pharmacy* 10 (82): 1–23, https://doi.org/10.3390/pharmacy10040082

Anderson, M. 1990. *Colour therapy. The application of color for healing, diagnosis and well-being*. Wellingborough: The Antiquarian Press (original 1975).

Batchelor, D. 2000. *Chromophobia*. London: Reaktion Books.

Benedictow, O. J. 2004. *The Black Death 1346–1353. The complete history*. Suffolk: The Boydell Press.

Braun, L. 2014. *Breathing race into the machine. The surprising career of the spirometer from plantation to genetics*. Minneapolis, MN: University of Minnesota Press.

Bruno, N., Martani, M. Corsini, C., and Oleari, C. 2013. The effect of the color red on consuming food does not depend on achromatic (Michelson) contrast and extends to rubbing cream on the skin. *Appetite* 71: 307–313, https://doi.org/10.1016/j.appet.2013.08.012

Dass, A. 2023. *Humanae*, https://angelicadass.com/photography/humanae

De Craen, A. J. M., Roos, P. J., De Vries, A. L., and Kleijnen, J. 1996. Effect of colour of drugs: systematic review of perceived effect of drugs and of their effectiveness. *British Medical Journal* 313 (7072): 1624–1626, https://doi.org/10.1136/bmj.313.7072.1624

Dyer, R. 2017. *White. Twentieth anniversary edition*. London and New York: Routledge, https://doi.org/10.4324/9781315544786

Gage, J. 1999. *Colour and meaning. Art, science and symbolism*. London: Thames & Hudson.

Gage, J. 2013. *Colour and culture. Practice and meaning from Antiquity to abstraction*. London: Thames & Hudson.

Garcia, M. A., Homan, P. A., Gracia, C., and Brown, T. H. 2021. The color of COVID-19: Structural racism and the disproportionate impact of the pandemic on older Black and Latinx adults. *The Journals of Gerontology. Series B, Psychological Sciences and Social Sciences* 76 (3): e75–e80, https://doi.org/10.1093/geronb/gbaa114

Guéguen, N., and Jacob, C. 2014. Clothing color and tipping: Gentlemen patrons give more tips to waitresses with red clothes. *Journal of Hospitality & Tourism Research* 38 (2): 275–280, https://doi.org/10.1177/1096348012442546

Haidt, J. 2001. The emotional dog and its rational tail: A social intuitionist approach to moral judgment. *Psychological Review* 198 (4): 814–834, https://doi.org/10.1037//0033-295x.108.4.814

Hopkins, D. R. 2002. *The greatest killer. Smallpox in history.* Chicago, IL and London: The University of Chicago Press.

Jablonski, N. G. 2012. *Living color. The biological and social meaning of skin color.* Berkeley, CA and London: University of California Press, https://doi.org/10.1525/9780520953772

Jones, R. 2021. *White borders. The history of race and immigration in the United States from Chinese exclusion to the border wall.* Boston, MA: Beacon Press.

Kwallek, N., Lewis, C. M., Lin-Hsiao, J. W. D., and Woodson, H. 1996. Effects of nine monochromatic office interior colors on clerical tasks and worker mood. *Color Research and Application* 21 (6): 448–458.

Levinas, E. 1961. *Totalité et Infini. Essai sur l'extériorité.* [Totality and Infinity. An essay on exteriority] Den Haag: Martinus Nijhoff.

Macneill, P. 2017. Balancing bioethics by sensing the aesthetic. *Bioethics* 31: 631–643, https://doi.org/10.1111/bioe.12390

Merleau-Ponty, M. 1964. *The primacy of perception: And other essays on phenomenological psychology, the philosophy of art, history, and politics.* Evanston, IL: Northwestern University Press.

Mesman, J. 2021. *Opgroeien in kleur. Opvoeden zonder vooroordelen.* [Growing up in color. Educating without prejudices] Amsterdam: Uitgeverij Balans.

Moller, A. C., Elliott, A. J., and Maier, M. A. 2009. Basic hue-meaning associations. *Emotion* 9 (6): 898–902, https://doi.org/10.1037/a0017811

Olguntürk, N., Aslanoğlu, R., and Ulusoy, B. 2021. Color in hospitals. In: Shamey, R. (ed.), *Encyclopedia of color science and technology.* Berlin and Heidelberg: Springer, 1–4, https://doi.org/10.1007/978-3-642-27851-8_449-1

Pantalony, D. 2009. The colour of medicine. *Canadian Medical Association Journal* 181 (6–7): 402–403, https://doi.org/10.1503/cmaj.091058

Pastoureau, M. 2010. *Les couleurs de nos souvenirs.* [The colors of our memories] Paris: Éditions du Seuil.

Pastoureau, M. 2014. *Green. The history of a color.* Princeton, NJ and Oxford: Princeton University Press.

Pastoureau, M. 2017. *Red. The history of a color.* Princeton, NJ and Oxford: Princeton University Press.

Pastoureau, M. 2019. *Yellow. The history of a color*. Princeton, NJ and Oxford: Princeton University Press.

Pastoureau, M., and Simonnet, D. 2005. *Le petit livre des couleurs*. [The little book of colors] Paris: Éditions du Panama.

Posth, C., Yu, H., Ghalichi, A. et al. 2023. Palaeogenomics of Upper Palaeolithic to Neolithic European hunter-gatherers. *Nature* 615: 117–126.

Riley, C. A. 1995. *Color codes. Modern theories of color in philosophy, painting and architecture, literature, music, and psychology*. Hanover and London: University Press of New England.

Romano, C. 2020. *De la couleur*. [About color] Paris: Gallimard.

Rose, S. 2022. A deadly ideology: How the "great replacement theory" went mainstream. *The Guardian*, 8 June, https://www.theguardian.com/world/2022/jun/08/a-deadly-ideology-how-the-great-replacement-theory-went-mainstream

Russell, C. 2022. Meeting the moment: Bioethics in the time of Black Lives Matter. *American Journal of Bioethics* 22 (3): 9–21, https://doi.org/10.1080/15265161.2021.2001093

Sallis, R. E., and Buckalew, L. W. 1984. Relation of capsule color and perceived potency. *Perceptual and Motor Skills* 58: 897–898.

Shendruk, A. 2021. These are the colorful and confusing Covid-19 alerts used around the world. *Yahoo News*, 23 April, https://qz.com/1996422/the-differences-between-color-coded-Covid-19-warnings-globally

Siegel, E. 2020. No, the COVID-19 coronavirus is not actually red. *Forbes*, 15 April, https://www.forbes.com/sites/startswithabang/2020/04/15/no-the-covid-19-coronavirus-is-not-actually-red/

Spence, C., Levitan, C. A., Shankar, M. U., and Zampini, M. 2010. Does food color influence taste and flavor perception in humans? *Chemosensory Perception* 3 (1), 68–84, https://doi.org/10.1007/s12078-010-9067-z

Starobinski, J. 1981. Chlorosis—the "green sickness". *Psychological Medicine* 11 (3): 459–468.

St Clair, K. 2016. *The secret lives of color*. London: Penguin.

Stokowski, L. A. 2011. Fundamentals of phototherapy for neonatal jaundice. *Advances in Neonatal Care* 11 (5 Suppl): S10–21, https://doi.org/10.1097/anc.0b013e31822ee62c

Street, B. 2018. *Art unfolded: A history of art in four colours*. Lewes: Ilex Press.

Taussig, M. 2009. *What color is sacred?* Chicago, IL and London: The University of Chicago Press.

Ulrich, R. S. 1984. View through a window may influence recovery from surgery. *Science* 224 (4647): 420–421.

Weyns, W. 2023. *Wie Wat Woke. Een cultuurkritische benadering van wat we (on) rechtvaardig vinden*. [Who What Woke. A cultural critique of what we find (un)just] Kalmthout: Pelckmans Uitgevers.

Wiercioch-Kuzianik, K., and Babel, P. 2019. Color hurts. The effect of color on pain perception. *Pain Medicine* 20 (10): 1955–1962, https://doi.org/10.1093/pm/pny285

Yu, L., Westland, S., Chen, Y., and Li, Z. 2021. Colour associations and consumer product-colour purchase decisions. *Color Research and Application* 46: 1119–1127, https://doi.org/10.1002/col.22659

2. The Nature of Color

2.1 Introduction

In Ancient Greek the word *pharmakon* in general refers to a tool, or means for or against something, and more specifically to a healing or deadly drug, poison, charm or spell, dye or color. The same person producing *pharmaka* therefore makes drugs to heal ailments and diseases but also pigments and paints. This is not surprising since many natural pigments were assumed to have medicinal properties and were used in recipes for medication. At the same time, ingredients could be healing but also poisonous. One of the most commonly applied pigments was lead white, already known in Antiquity. It is basic lead carbonate, prepared from a natural mineral, and used as paint but also cosmetic in various cultures (St Clair 2016). It is highly toxic for those manufacturing it but also for people who use it to whiten their skin. The white color it produces is seen as beautiful, charming and seducing but also intoxicating and dangerous.

As a pharmakon, color is primarily regarded as a material that covers things and beings. Color is like an envelope, a second skin that encompasses an object or body. This conception is reflected in Indo-European languages where the Latin word *color* is derived from the verb *celare*, which means to conceal. In Ancient Greek, the word for color (*khroma*) refers to skin or bodily surface (Pastoureau 2019). The etymology of the term and the ambivalence of color as matter has instigated many philosophers to reflect on the nature of color.

In our experience, colors are real and out there in the world. At the same time they are variable; depending on light, context and point of view we see different colors and not all observers see the same colors. To illustrate the color blue we may point to the sky, but the color of the sky changes rapidly during twilight. This changeability makes it doubtful whether the color that an object appears to have is really a property of the object or rather dependent on the perceiver. Is it true that color, as expressed in the origin of the term, is deceptive and not really part of the world as we see it? Consequently, the issue of color refers to a broader philosophical concern about the certainty of knowledge and the reliability of our senses. It also problematizes the status of scientific understanding. Science tells us that only physical facts exist in the world, and that colors should be identified with physical properties such as wavelength. When we look at the grass it looks green but from the point of view of science it has no color. Our visual experience is misleading us. But if it is an illusion that the world which we experience with our sensory organs is colored, the same is true not only for other qualitative properties such as smell, sound and taste, but also for concepts that are crucial to human existence, like agency, mind, personhood, responsibility and moral values. The subject of color has attracted the attention of philosophers since it relates to a range of philosophical issues concerning appearance and reality, subjectivity and objectivity, science and common sense, mind and matter.

In this chapter, a brief description will be provided of the main philosophical theories of color. Realism refers to a set of theories that regard color as an objective property of objects in the outside world which is independent from the perceiving subjects. Antirealist theories postulate that colors are products of the visual system and that they do not exist without perceiving subjects. Because it is typical for the experience of color to be not merely located in the outside world or in the mind, but situated between the objective and the subjective world, a third type of theory emphasizes color relationism. These theories assume that color vision is not simply detection of color but has a specific purpose, namely identification of features of the environment that are useful for the perceiving organism, helping it to explore the world and to flourish and survive. However, as articulated in phenomenological philosophy, the world is not the objective one of the natural sciences but it is the world as directly experienced in everyday life, and that is prior to scientific analyses and reflection. Human beings are part of this life-world, and they interact with it through perceptual activity. In this world, colors do not exist alone but always as colors of certain objects, operating within a visual field. Through colors, objects have meaning and significance, and that is why colors move us.

2.2 The Traditional View

In the *Timaeus*, Plato argues that colors are effluences emanating from bodies in the surrounding world. These bodies emit a stream of particles of fire that are perceived as colors. The color of a body is independent of the sentient beings which perceive it. Therefore colors are real and objective. In other works, however, Plato describes color as the private object of experience of the individual observer; it exists as long as a particular observer interacts with a particular object (Ierodiakonou 2005). Thus the status of color as a pharmakon is not clear: is it real and existing in the world or is it just how things in the world appear since, in fact, material objects have no color, as argued by Democritus and Lucretius, and color is merely a subjective experience? For a long time, these questions were not relevant. The received view on color until the seventeenth century was Aristotelian. It accepts that colors exist in the world, and that they reside in objects. If we see a red flower, it is red. According to Aristotle, colors are there even if they are not visible and if there are no sentient beings to perceive them. Light is required to make them actually perceived. Various colors are the result of the mixture of two basic colors, white and black (Ierodiakonou 2018).

2.3 The Scientific Revolution

Since the emergence of natural sciences in the seventeenth century, reality has been interpreted as quantitative, material and mechanistic. Qualitative properties of objects such as colors become problematic. They can only be incorporated into the scientific view of the world as constituted by physical components if they can be explained in terms of physics. The experiments of Isaac Newton in the 1660s were crucial for the physical interpretation of color.

Fig. 2.1 Isaac Newton's prism experiment. Image created by Castellsferran (2020), Wikimedia, https://commons.wikimedia.org/wiki/File:Experiment_dels_primes_d'Isaac_Newton_-_Refracci%C3%B3_de_la_llum.png#/media/File:Experiment_dels_primes_d'Isaac_Newton_-_Refracci%C3%B3_de_la_llum.png, CC BY-SA 4.0.

Newton relates color to light, following the suggestion of Aristotle that light is the activator of color (Gage 1999). When visible light is refracted through a prism, it contains seven fundamental or spectral colors (red, orange, yellow, green, blue, indigo and violet). Each color has a specific length and frequency of electromagnetic waves. Red has the longest wavelength and violet the shortest. The surfaces of objects in the world absorb and reflect these waves so that specific colors become visible. The chemical constitution of materials determines what wavelengths are absorbed and reflected. If all wavelengths are reflected, we see white; if they are all absorbed, we see black. If the longest wavelengths are reflected, and the others absorbed, we see red. Newton's studies have promoted physicalistic theories: colors are physical properties of material bodies and entities, and can be measured since they are written in the language of mathematics (Romano 2020, 66).

Fig. 2.2 Color wheel wavelengths. Image created by Amousey (2023), Wikimedia, https://commons.wikimedia.org/wiki/File:Color_wheel_vector.svg#/media/File:Color_wheel_vector.svg, CC BY-SA 4.0.

2.4 Primary and Secondary Qualities

The view that color is not a property of things in the world also became popular among philosophers. René Descartes (1596–1650) compares color with pain. When we burn our hand in a fire, the pain is not in the flames nor in our hand but it exists in our mind; similarly it is an error to assume that colors exist outside us. Such error is the result of conditioning since early childhood because we have learned to associate sensations and perceptions with real things outside our mind (Descartes 1644). Colors are therefore problematic since they seduce and mislead us into thinking that they exist in the outside world. They do not really belong to the natural world, as described and explained by science. Colors are appearance, ideas within us, and not properties of things whereas physical properties of things are the cause of our perception of colors. In the history of philosophy, the status of color is defined by a classical distinction made by John Locke (1632–1704) between primary and secondary qualities. Colors are qualities and properties that require a subject, a substance; they cannot exist by themselves. According to Locke, qualities come in two types: primary qualities (such as bulk, number and motion of particles) which are essential and intrinsic; they define a substance. Secondary qualities are accidental or contingent; they are powers (or dispositions) to produce ideas in perceivers caused by the particular constitution of the primary qualities of the substance (Locke 1975). The temptation is to imagine that colors, as ideas in our mind produced by the primary quality of objects, are in fact in or attached to the objects themselves. Following Descartes, Locke repeats that this is a misunderstanding. Milk seems white but it is not; because of its specific constitution it has a power to produce in us the idea of white, but the color itself is not in the outside world.

2.5 Philosophies of Color

Color is the subject of numerous philosophical studies and debates, often in connection to the findings of contemporary science of color. It is not the aim of this chapter to analyze or summarize current philosophies of color but it will highlight several main issues which are important to understand the role, significance and use of color in healthcare and bioethics. Typical for many philosophies is that they elucidate the nature of color by localizing it either in the world or in the mind. In contrast to the traditional view they assume that there is a strict dualism between the subjective world of sensation and perception, and the objective world of physical facts (Chirimuuta 2015).

Realist theories, such as physicalism, identify colors with the physical causes of visual experiences of colors. They proceed on the basis of the commonsense view that colors are intrinsic, perceiver-independent properties of objects while they reconcile this view with scientific findings through interpreting colors as physical, and thus objective, properties (for example wavelengths of light) (Smart 1997). Doing so, it is not clear whether the nature of color as encountered in everyday life is indeed clarified. The question of color is answered with a physical definition, but it is a problem how light rays are able to cause sensations of color. Identifying colors with physical states has the effect that colors disappear from the world in which they are experienced, undergone and enjoyed

(Thompson 1995, 32). There is often a clash between physical reality and perception of color. Depending on light, weather and context, colors are variable and seem to change whereas the physical causes remain the same. This also applies to the phenomenon of color constancy: perceived colors are relatively stable under changing spectral lighting conditions. Similar discrepancies occurred in Newton's work. He identified seven spectral colors because he wanted to find a correspondence with the main tones in the musical scale. Visual experience, however, is different. Many people cannot differentiate indigo from blue (Eckstut and Echstut 2013). In the history of art, five colors have been regarded as primary (black, white, red, blue and yellow). This was based on pigments not light; only a few colors are needed to mix and produce other colors. In Newton's system, black and white are not colors at all.

Other theories reject the idea that colors are objective properties of things. They advocate antirealism (Chirimuuta 2015): physical descriptions of the world do not refer to colors, and the appearance of color is therefore an illusion. Hardin (1993, 111), for example, concludes that "…we have no good reason to believe that there are colored objects." These theories are often based on neurophysiological studies. They emphasize the importance of perception: wavelengths of visible light are only colors when we see them. Colors are just names or labels, and without perceiving subjects they are not there. They are produced and constructed in the visual system and do not exist outside of perception. The retina has two types of photosensitive cells: rods (to distinguish between dark and light) and cones (to distinguish among colors). There are three types of cones, sensitive to short, medium or long wavelengths, and usually labeled as blue, green or red. Light waves activate the rods and cones which then send messages to the visual cortex of the brain that interprets the physical sensation as the perception of colors, depending on what kind of cones are activated. Without cones wavelengths have no color. Neurophysiological theories thus provide an explanation of color in anatomical and physiological terms but they are subjective theories in the sense that color is not a property of objects in the outside world but completely perceived and produced within the perceiving subject. These theories reiterate Descartes's view that colors are mental constructs, ideas within us, while we have the tendency to project them on external objects. The idea of projection is famously underlined by

David Hume (1711–1776) who compares moral notions of good and evil, virtue and vice with colors (and sounds, heat and cold) which are not qualities in objects but perceptions of the mind (Hume 1978). They are projected on an achromatic world just like paint is used to cover and embellish objects. Arthur Schopenhauer (1788–1860), to name another philosopher, points out that colors are constituted through retinal and cerebral activities. He concludes that color is purely subjective (Schopenhauer 1994). But when colors are only situated in the eyes and central nervous system, they are in fact private mental states. How can we be sure that we see the same colors as our neighbors when there is no connection between wavelengths of light, coming from the outside world, and the neuronal activities within our brain that produce color, and no criteria to distinguish perception from illusion?

2.6 Color Relationism

One conclusion so far is that objective or realist as well as subjective or antirealist interpretations of color have the same effect: they assign a particular location to color; it is outside in the physical world or inside in our brains or mind (Romano 2020, 135). They reduce colors to either sensations and sense data that stimulate the human visual system or to perceptions that are constructed in the visual system itself. According to realism and antirealism, milk itself is not white. Because its molecular composition reflects all wavelengths of the visible light, our visual experience of milk is white. Since the wavelengths reflected by the milk activate the neuronal system, we have the perception of whiteness. In both cases the color does not exist in the outside world unless we take the physicalist view that identifies colors with wavelengths of light. We experience a significant discrepancy: the reality experienced through our eyes is colored whereas the reality according to physics and neurophysiology has no color (Chirimuuta 2015, 8).

To overcome the divergence of science and experience, it is helpful to critically examine some assumptions that underly the philosophical views of color. Chirimuuta (2015) mentions first of all the detection model of perception. Both realism and antirealism assume that through sensory perception true knowledge of the world is obtained. With the senses, physical properties of objects in the world can be detected. The eyes in particular are windows onto the world, comparable to physical measuring devices. The second assumption is reification of color. This is the idea that color can be abstracted not only from the context of the material world but also from other visual qualities such as size and shape. Color is indeed an entity or thing like dye or paint that covers an achromatic world, and that may be detected with perception. If it is real, the visual system will uncover it; if it is not real, it is the mind that paints the surrounding world. Color is therefore a quality independent from other secondary qualities that can be separated from primary qualities or rather the quantitative constituents of the world. The third assumption is that colors are located since there is a strict demarcation between the outside world of physics and the inside neuronal world of the mind or the brain. The consequence of these three assumptions is that color is not regarded as the interaction between perceiver and object. Together, these assumptions make it impossible to conclude, as Chirimuuta (2015, 17) underlines, that color vision is "a joint product of the perceiver and perceived, so colors are relational in this sense."

Usually, colors are regarded as sensations or perceptions rather than as experiences which relate objective and subjective elements. Color perception is the detection of the chromatic properties of objects, separable from other visual aspects. Color is like an add-on that is dispensable without affecting the visual system. However, if color vision cannot be separated from the rest of vision, and is understood as interaction between perceiver and environment, colors do not simply belong to objects or observers. This idea of relation is expressed in the theory that colors are dispositions, for example in the work of Locke (1975, 135 ff). According to him, colors such as red or white are not in the objects themselves but their perception is produced by the powers or dispositions of substances. At the same time, these dispositions are the result of intrinsic qualities of substances, i.e. their atomic and molecular structure. Objects perceived as red or white have the disposition to produce the perception of these colors because of a physical infrastructure that absorbs and transmits certain wavelengths of light which reach the eye and are decoded in the central nervous system so that a color is perceived. Colors are real, but since they produce sensations they are subjective as ideas in the mind (the idea of redness or the idea of whiteness). They are therefore objective as well as subjective (Romano

2020, 40, 60 ff). However, precisely because of this duality, the notion of disposition is criticized. It is argued that dispositions do not cause their manifestations (Jackson and Pargetter 1997). For example, when a glass is dropped, the cause of its breaking is not its fragility but its internal constitution. Likewise, when an object looks red it is not because it has a particular disposition but rather because of a specific physical constitution, and then what really matters are physicalistic theories.

2.7 Ecological Theories

The view of color as a relational property is elaborated in ecological theories (Thompson 1995). Colors have a dual nature. They are Janus-faced, with one face turned to the world, and the other to the perceiving subject. The implication is that colors should not be interpreted as solely an intrinsic property of an object (determined by its physical characteristics) nor as an idea or mental construct (determined at neuronal level) but more appropriately as a relational property that brings together the object and its environment as well as the perceivers. Colors have a reality in the phenomenal world which is partly independent of human perceivers, and it is also more than a private mental state in the perceivers. This relational theory is influenced by an ecological view of colors that attributes specific functions to perception. The aim of perception is to identify certain characteristics of the environment that are useful for the survival of a species, to discriminate between beneficial and harmful objects, and to classify them. Non-mammals generally have a richer capacity for color vision than mammals. For example, many birds, particularly diurnal birds, have color vision in the near-ultraviolet region. This ultraviolet sensitivity is helpful to facilitate aerial navigation but also to recognize fruit trees and to identify ripe fruits among green vegetation. The same is true for forager bees; their sensitivity for ultraviolet light helps them to associate nectar or pollen with color patterns of flowers (Thompson 1995, 172). Color vision has various functions, for example perception of depth and distance, shape, contours, texture: "color perception is part and parcel of the mechanism for perceiving numerous different properties of objects…" (Chirimuuta 2015, 128). It is also necessary to identify objects (such as edible food); and furthermore has cognitive functions—for example, it helps to memorize objects and distinguish material differences between them.

The emphasis is therefore on the usefulness of perception. Seeing with colors allows the perceiver to undertake activities; it is active interaction and engagement with the world, depending on the interests and needs of the perceiver. From an ecological perspective, color vision is primarily interaction with the outside world, not merely a matter of observation. It helps living beings to stay alive in their biological environments and to explore these environments (Romano 2020, 137). In the non-human world, seeing colors not only helps to detect objects against their backgrounds, but also guides interactions and behaviors because it serves to identify organisms belonging to the same species and to apprehend their sexual state. A recently studied example are desert locusts; when they live alone they are usually green or sand-colored to camouflage them against their background, but in large swarms adult males turn yellow. This color is a warning signal that prevents sexual harassment from other males (Cullen et al. 2022). In the animal kingdom yellow is often a color that adapts animals to their habitat (e.g. the giraffe and lion) but it can also be a sign of danger (e.g. yellow frogs that are poisonous).

Fig. 2.3 Yellow-banded poison dart frog. Photo by Holger Krisp (2013), Wikimedia, https://commons.wikimedia.org/wiki/File:Bumblebee_Poison_Frog_Dendrobates_leucomelas.jpg#/media/File:Bumblebee_Poison_Frog_Dendrobates_leucomelas.jpg, CC BY 3.0.

For human beings colors have similar functional roles, particularly within social and cultural contexts. Identifying colors has an instrumental value: "… it is rarely helpful to know the colour of a thing simply for its own sake: an interest in colour is typically serving some further end of the organism" (Campbell 2017, 185). Colors are used to distinguish and classify a specific social status, rank or profession. In Roman Antiquity, for example, red is the color of power and only preserved for specified officials such as generals and consuls. In many societies the use of color in clothing and food has been strictly regulated. Colors also have specific functions in religious contexts, such as referring to special days in Catholic liturgy. Monasterial orders have different colors for the habits of monks, while Buddhist monks wear orange. Colors are also used to diagnose diseases, and later, to determine microorganisms in the laboratory. Paints and pigments are applied to embellish the body with various types of make-up. On the basis of skin color, people are often classified and ranked.

2.8 Color Adverbialism

Dispositional and ecological theories consider visual experiences as revealing the nature of color in the outside world. Talking about color means pointing to the world around us. Both theories are relational since they attempt to bridge the gap between inner and outer, but Chirimuuta criticizes them because they connect the outerness of color with an ontological commitment that colors are properties instantiated by external objects (Chirimuuta 2015, 51). In her view they still regard color vision as the detection of mind-independent properties, and color as a property separable from other visual features. It is insufficiently expressed that colors are in between the physical and the mental, and that they are not locatable as inner or outer. It is necessary to go beyond the dualism of object and subject, and to avoid reification of color. According to Chirimuuta (2015) color vision is not first of all a way of seeing colors (with perception conceived as detection) but a way of seeing things (with perception conceived as discrimination). Colors are properties of specific kinds of events or processes, i.e. perceptual interactions. From this perspective, color is not situated in the world (realism), in the mind (antirealism), or in both simultaneously (relationism). Instead, color exists solely as a property of a perceptual process, and thus is better described adverbially rather than as a substantive. However, it is objected that in visual experience, as well as in the way colors are predicated in natural languages, colors qualify individuals rather than events. Colors are typically perceived as properties of things in a visual context (Thompson 1995). Color adverbialism would also counterintuitively imply that without perceptual events (for example when there are no perceivers, or when they close their eyes) colors no longer remain in place (Cohen 2015).

2.9 The Phenomenological Perspective

The basic idea that humans are relational beings embedded in the surrounding world is crucial in the philosophy of phenomenology. It emphasizes that the analysis of phenomena like color should not start from the assumption of an objective world, and also not from a pure, constituting consciousness but from their unity and fusion. Being human means being-in-the-world, embodied being in a situation. It is impossible to separate the living organism from the world with which it interacts, but equally impossible to detach the surrounding world from the organism. However, this world is not the objective, physical or biological world of the natural sciences but the world encountered in everyday life and revealed in immediate, lived experience (Spurling 1977). In the immediacy of the experience there is no distance between humans and world; being human is fused with the world. We are, live and act in the life-world (*le monde vécu*) before we make this world the object of scientific reflection and analysis. The cognitive relationship between knowing subject and known object is preceded by the intertwining of human beings and world, and is first of all a perceptive relationship. Perception takes place at a pre-reflective level; it brings us into contact with the world that is prior to scientific knowledge; it is according to Maurice Merleau-Ponty (1908–1961), in contrast to knowing, a living communication with the world that makes it present to us as the familiar place of our life (Merleau-Ponty 1945, 64–65).

Perception is an activity or process that connects perceivers to their environments, as is also pointed out in other relational theories of color. For Merleau-Ponty, perception is not purely physiological nor psychological or reflective but a pre-reflective and pre-conscious interaction with the life-world. This materializes because we are embodied beings; our body is the only viewpoint on the world that we have. The body is situated in and part of the world and, at the same time, open to it. It is the body that actively organizes and patterns the visual field, depending on the tasks and interests of the organism, and determining possible areas of activity. Perception therefore is a bodily event, preceding reflecting and analyzing consciousness. The world is presented to me through my body.

Perception furthermore refers to context; for example, how and whether colors are perceived depends on the level and degree of lighting. We also do not perceive discrete, isolated objects but they are always linked to and surrounded by others, part of a 'field' and presented within a horizon. Colors are never simply colors; they are colors of specific objects. The blue of a carpet is not the same as the blue of wool. Colors form a system, embedded in objects and interwoven with other sensory data within a visual or phenomenal field. It is not possible to describe a color without referring to the object it belongs to—whether a carpet or woolen clothing—which feels and smells different, has a different weight, and responds uniquely to sound. The significance of perceived objects is immediately grasped in perception. According to Merleau-Ponty, a color is felt and the body responds before we are even aware that we see it. Colors have *signification motrice*: they touch and move us (Merleau-Ponty 1945, 243 ff). Red may exaggerate my reactions without me noticing, while green is restful. Colors bring about certain attitudes of the body and a particular way of perceiving, but this should not be construed as the effect of a specific quality upon an observing subject but as a pre-personal experience in the visual or phenomenal field. This field is meaningfully structured by the phenomenal body which is oriented towards projects and intentions calling for activities in the world. Phenomena of color are not in our minds or brains but they are how things present themselves to us in the life-world. They are immediately apprehended as meaningful and charged with significance. Since perception is communication or communion, they have the power to move us: "Quality, light, colour, depth, which are there before us, are there only because they awaken an echo in our body and the body welcomes them" (Merleau-Ponty 1964, 164).

2.10 Conclusion

In this chapter various philosophies of color have been discussed. It is common to distinguish between realist theories that posit that colors are objective properties of things, and antirealist theories that consider colors as ideas in the mind or mental constructs. Both types of theories are strongly influenced by scientific findings, particularly in physics, optics and neurophysiology. Consequently, philosophical theories often share the same assumptions as scientific theories. Colors are regarded as qualitative properties that have no proper place in the scientific image of the world as constituted by material and quantitative components; they are either physical or non-physical phenomena. They must be located in the outside world or inside the mind or the brain. But in both cases, color as experienced in everyday life disappears: as an objective, perceiver-independent property it is the effect of the physical state of things, and as a subjective, perceiver-dependent property it is produced by the visual system and erroneously projected on external objects. What the eyes perceive as colored has no color in the reality of natural sciences. Perception in these theories is detection and recording of objects in the outside world.

Color and its perception is understood differently in a third type of philosophical theories that emphasize color vision as the interaction between environment and perceiver. These relationist theories assume that color cannot be separated from other visual aspects of objects (such as form and matter). They also point out that perception is not merely detection but has a range of functions, specifically discriminatory and cognitive ones. From an evolutionary and ecological perspective color vision helps organisms to explore their environment, and to structure it according to what is relevant and helpful for their existence and survival.

Color as a relational experience is most clearly articulated in phenomenological philosophy. It stresses that perception is action, a process of connecting perceivers with their environment in order to guide behavior. Colors therefore cannot be located inside or outside the organism. Perceiving is an act not of the brain but of the organism; it is "a particular sort of activity in which the subject engages..." (Thompson 1995, 298). Furthermore, it is an embodied activity. The embodied organism is the subject of perceptual experience. The life-world of the organism is structured and patterned into relevant visual fields through the body. This highlights the importance of meaning. Color is best understood, in the words of Chirimuuta as "a means by which organisms make sense of the complexity of the natural environment" (Chirimuuta 2015, 15). Finally, relationism in general, and phenomenology in particular, emphasize the importance of context and life-world. As the world is not simply the physical, mechanistic and material world of the natural sciences, independent of the organism, it is impossible to separate the organism from the surroundings which with it interacts. At the same time, the organism is not a passive spectator that receives sensations from physical impressions but should be regarded as an active and exploring being making sense of its environment. In the life-world, colors do not exist on their own as abstract entities but they are always colors of certain objects and are operating within a visual field. In perceptual activity they highlight certain aspects of this field because they point to their significance and meaning. According to this philosophical perspective, seeing colors is not a neutral observation but has the power to touch and move us. This power will be the subject of the following chapter.

References

Campbell, J. 1997. A simple view of colour. In: Byrne, A., and Hilbert, D. R. (eds), *Readings on color. Volume I: The philosophy of color*. Cambridge, MA and London: The MIT Press, 177–190.

Chirimuuta, M. 2015. *Outside color. Perceptual science and the puzzle of color in philosophy*. Cambridge, MA and London: The MIT Press, https://doi.org/10.7551/mitpress/9780262029087.001.0001

Cohen, J. 2015. Outside color: Perceptual science and the puzzle of color in philosophy [review]. *Notre Dame Philosophical Reviews*, 21 October, https://ndpr.nd.edu/reviews/outside-color-perceptual-science-and-the-puzzle-of-color-in-philosophy

Cullen, D. A., Sword, G. A., Rosenthal, G. G. et al. 2022. Sexual repurposing of juvenile aposematism in locusts. *Proceedings of the National Academy of Sciences of the United States of America* 119 (34): e2200759119, https://doi.org/10.1073/pnas.2200759119

Descartes, R. 1644. *Principia philosophiae*. Amsterdam: Ludovicum Elzevirium.

Eckstut, J., and Eckstut, A. 2013. *The secret language of color*. New York: Black Dog & Leventhal Publishers.

Gage, J. 1999. *Colour and meaning. Art, science and symbolism*. London: Thames & Hudson.

Hardin, C. L. 1993. *Color for philosophers. Unweaving the rainbow*. Indianapolis, IN: Hackett Publishing Company (2nd edition).

Hume, D. 1978. *A treatise of human nature*. Edited by L. A. Selby-Bigge and revised by P. H. Nidditch. Oxford: Clarendon Press (2nd edition, original 1739).

Ierodiakonou K. 2005. Plato's theory on colors in the Timaeus. *Rhizai. A Journal for Ancient Philosophy and Science* 2: 219–233.

Ierodiakonou, K. 2018. Aristotle and Alexander of Aphrodisias on colour. In: B. Byden and F. Radovic (eds), *The Parva naturalia in Greek, Arabic and Latin Aristotelianism*, Studies in the History of Philosophy of Mind 17. Cham: Springer, 77–90, https://doi.org/10.1007/978-3-319-26904-7_4

Jackson, F. and Pargetter, R. 1997. An objectivist's guide to subjectivism about color. In: Byrne, A., and Hilbert, D. R. (eds), *Readings on color. Volume I: The philosophy of color*. Cambridge, MA and London: The MIT Press, 67–79.

Locke, J. 1975. *An essay concerning human understanding*. Edited by P. H. Nidditch. Oxford: Oxford University Press (original 1690).

Merleau-Ponty, M. 1945. *Phénoménologie de la perception*. [Phenomenology of perception] Paris: Gallimard.

Merleau-Ponty, M. 1964. *The primacy of perception: And other essays on phenomenological psychology, the philosophy of art, history, and politics*. Evanston, IL: Northwestern University Press.

Pastoureau, M. 2019. *Yellow. The history of a color*. Princeton, NJ and Oxford: Princeton University Press.

Romano, C. 2020. *De la couleur*. [About color] Paris: Gallimard.

Schopenhauer, A. 1994. *On vision and colours*. Edited by D. E. Cartwright and translated by E. F. J. Payne. London: Bloomsbury Publishers (original 1816).

Smart, J. J. C. 1997. On some criticisms of a physicalist theory of colors. In: Byrne, A., and Hilbert, D. R. (eds), *Readings on color. Volume I: The philosophy of color*. Cambridge, MA and London: The MIT Press, 1–10.

Spurling, L. 1977. *Phenomenology and the social world*. London: Routledge & Kegan Paul.

St Clair, K. 2016. *The secret lives of color*. London: Penguin.

Thompson, E. 1995. *Colour vision. A study in cognitive science and the philosophy of perception*. London and New York: Routledge.

3. The Power of Color

3.1 Introduction

Color is often presented as a curious and enigmatic phenomenon. It is studied from a range of perspectives and disciplines: physics, physiology, neurobiology, medicine, psychology, sociology, cultural sciences, linguistics and aesthetics, frequently without reference to implied philosophical perspectives on the nature of color. Nowadays, colors play a major practical role in a wide variety of activities such as art, architecture, painting, decorating, fashion, cosmetics, clothing, advertising, branding, gardening, ceremonies and traffic (Tan et al. 2011). This practical role is also a product of the growth of a major manufacturing industry: in the nineteenth century, the development of chemistry allowed the production of synthetic colorants. Also the application of color has become commercialized with the emergence of a special profession of color consultants, united in professional bodies such as IACC (International Association of Color Consultants/Designers). They advise about the use of colors in architecture, interior and environmental design, psychology and marketing. Given this background, many studies on color have a practical intent: they want to investigate how colors affect and influence human behavior. They assume that color is not inert but exercises particular powers. Without much theorizing about what color essentially is, they consider it as it is experienced, and then focus on its role in human interactions. In these practical approaches, color is not interpreted as simply a surface or ornament but as something that can transmit and communicate codes, feelings, attitudes, prejudices and normative judgments. Color operates as a mover that influences language, behavior, imagination, and sentiment (Pastoureau and Simonnet 2005). These effects are particularly communicated by artists. For Henri Matisse (1869–1954), for example, colors are potent; he uses them in his paintings, not to transcribe nature but to express emotions. According to Wassily Kandinsky (1866–1944), color is a power which influences the soul (Riley 1995, 136, 142). It is a language without words, directly addressing our emotions and feelings (Street 2018). Although these practical applications assume that colors are powerful, they generally do not elaborate on the nature of color, with the exception of some painters. The focus is mostly on what color does, not on what it is. But as will be clear in the subsequent text, a relational perspective is frequently presupposed: colors are situated between the perceiving subject and the world, the interaction of which gives rise to emotions, feelings, attitudes and judgments.

This chapter will concentrate on the effects of color, especially on human behavior and emotions. First, it will examine how we talk about color. There is a curious discrepancy between the number of colors that can be discerned and the number of color names used in conversation. It is argued that under optimal conditions a trained normal observer can discriminate approximately ten million colors (Hardin 1993, 182). However, in everyday languages the number of names is much smaller. In English there are 7,500 color names, although some languages have only a few names. Popular web browsers such as Safari and Chrome support the use of 147 color names. One explanation of the discrepancy between seeing and naming is the difference between color matching in the laboratory and everyday life. Under normal conditions it is difficult to compare colors because of differences in light, shadow, contrast and materials (Hardin 1993). Another explanation is that the discrepancy is misleading because reference is made to different dimensions of color. Usually, color models make a distinction between hue, saturation (purity, intensity or chroma) and lightness (brightness, value or tone). Hue is the actual color such as red or yellow. Humans can distinguish two hundred steps from yellow through green and blue, and finally red. Saturation refers to the proportion of hue in a given color, its purity, determining how vivid or intense it is. For any hue, at least twenty steps can be distinguished. Lightness refers to how dark or light a color is, varying from white (full lightness) to black (no lightness). Humans can distinguish five hundred steps in this dimension (Thompson 1995). When the effects of colors are studied and compared in an experimental setting it is therefore necessary to specify the various dimensions. In normal daily circumstances colors can, however, be described in terms of a much more limited number of categories to which they belong, depending on the practical purposes of color vision. Perception of color is structured around basic or primary colors that are distinct and identifiable, although their number is contested (according to Empedocles and Plinius there are four basic colors; Aristotle distinguished five; Leonardo da Vinci and Isaac Newton, seven). Due to its categorical nature, color vision groups perceptual stimuli into hue categories that have significance and can guide behavior in various ways (Thompson 1995, 196).

This last explanation illustrates another problem that is often highlighted in color studies. If color perception in daily life is focused on a limited number of categories, why do languages generally have many more names for colors, even if they are significantly fewer than the colors that science can identify in laboratory conditions? Since we mostly communicate about colors in language, how do we know that when we call an object red, people with other languages see the same color? For a long time, it was assumed that the Ancient Greeks were not able to distinguish blue and green since they only had one word for both colors. It was also thought that there were hundreds of Inuit words for snow and nuances of white. From these observations it was concluded that color experience is ordered and classified by language, and thus must be arbitrary since there are different languages and cultures. According to the linguistic relativism of the Sapir-Whorf hypothesis, different languages do not represent the same world. This hypothesis, however, was undermined by research showing that across a variety of unrelated languages there is a correspondence of color categories. Berlin and Kay (1991) in particular demonstrated that in numerous languages there are, at most, eleven basic color terms, so that color categories are in fact universals.

After exploring the problem of the language of color, the chapter will examine the impact of color on emotions and feelings. Many studies propose that colors can evoke positive and negative emotional responses. The following section focuses on the effect of colors on human performance and behavior. Tests indicate, for example, that red has a negative influence on intellectual activity since it impairs concentration, while blue enhances the performance of creative tasks. Research on the effect of colors generally has a practical purpose which will be discussed in the following section of this chapter. If color or the perception of color influences human emotions and behavior, this finding has especial interest for the marketing industry, since it is assumed that decisions to buy products are not only based on brand, price and quality but also on color. These studies are furthermore relevant in the food industry. If particular colors modify the taste and flavor of food, the color of food packaging and food itself will influence consumer decisions and have an impact on appetite. That various effects of color have been observed in different fields is relevant for our later enquiry into the power of color in healthcare and bioethics.

3.2 Color Language

Language is not merely an instrument of communication but rather a tool to shape and organize our experiences and to create a coherent image of the world: "...the world is presented in a kaleidoscopic flux of impressions which has to be organised by our minds – and this means largely by the linguistic systems in our minds" (Whorf 1940, 3). According to the Sapir-Whorf hypothesis of linguistic relativity, the perceived reality is structured and classified depending on the vocabulary of a language and the needs and circumstances of the particular language community. Every language has a specific structure which determines how this community thinks and knows, and therefore generates a special perspective on the world that differs from other language communities. Color perception is not dependent on what we see but on the language we speak: it thus depends on the words with which we distinguish colors and ascribe them to objects and entities in the world. The assumption is that colors are located in the mind or brain and are communicated through language. There is an enormous variety of colors that can be named, and that can be ordered in identifiable categories. But each language does that in its own way, according to its grammar and logic, and depending on cultural, social and historical conventions about what is meaningful. A classic example is Inuit languages that use numerous different words for snow, so that the world is perceived in a different way with many more grades and types of snow than in other languages

(Whorf 1940, 6). Another example is that some languages (e.g. Greek and Russian) have two words for blue, one used for darker shades, the other for lighter shades. When people learn a new language, they obtain an ability to interpret the world differently, and to see different colors.

The hypothesis that color vocabulary determines what we are able to perceive is connected to an antirealist view of the nature of color: it is not actually out there in the world but the product of the human mind; color is a sensation, a private mental state. The above quotation from Whorf clarifies that when we perceive the world, we are confronted with a patchwork of sense data. The role of language is to create order, which is arbitrary and varies from one language to another. The electromagnetic spectrum presents a continuum from red to violet but the division of this continuum into specific colors is not determined by the physics of light but is a matter of name-giving and linguistic classification. This view of color and perception has been criticized from the perspective of relationism and phenomenology (Romano 2020). Colors are not perceived as simple, isolated and abstracted qualities, while the experience of colors is not amorphous and disordered. Perception is an activity that distinguishes categories of colors in connection to other visual aspects of objects. Color vision is structured around basic colors, even if their number is contested; its primary role is "to generate a relatively stable set of perceptual categories that can facilitate object identification and then guide behaviour accordingly" (Thompson 1995, 196). Colors are also not perceived as separate sense data but as exhibiting relations among themselves, such as complementarity and contrast: for example, the opponent relations between red and green, and yellow and blue.

Moreover, empirical studies have undermined the hypothesis of linguistic relativity. Berlin and Kay (1991) investigated more than ninety different languages and found that there are, at most, eleven basic color terms. Speakers of these languages could always identify the best example of a color category, although they disagreed about the boundaries between categories. There is not an arbitrary division of the color spectrum according to language but a pattern in all studied languages that revolves around basic colors: white, black, blue, yellow, green and red. Additionally, it was shown that the richness of the Inuit languages in regard to snow is a myth; these languages have more or less the same number of words for snow as English or French (Pullum 1991). Berlin and Kay conclude that basic color categories are "pan-human perceptual universals" (Berlin and Kay 1991, 109). Colors present an order or logic that is not a projection of the mind nor a construction of language (Romano 2020). The partition of the spectrum is not due to an arbitrary imposition of a category system but the effect of "universal constraints on the naming of colors" (Kay 2005).

Furthermore, Berlin and Kay found that different languages do not have the same number of basic color terms. There are languages with only two color words, while others distinguish eleven categories as basic. The authors postulate that the development of basic color categories follows a particular order in the evolution of a language. If languages have only two color words, then these are black and white; if they have three words, they are black, white and red. In following stages, more color terms are subsequently incorporated (green, yellow, blue and brown), whilst in the final stage, four more colors are encoded (purple, pink, orange and grey), bringing the total number of color categories to eleven. The core of color categories are four unique hues (red, green, yellow and blue) but it is not clear why they develop in this particular order. Hardin (1993) suggests that basic linguistic categories are based on common biological structures, while Berlin and Kay (1991) conjecture that the complexity of color vocabulary is related to cultural complexity and technological development since all languages of highly industrialized nations are in the final stage of the evolution of the color lexicon. If it is right that there are color categories common to all languages, it does not follow that these categories can be explained on the basis of biological or neurophysiological universals. The commonality can equally well be explained with reference to cultural and social universals (Maund 1995, Romano 2020).

While the universalistic approach to color language has undermined the hypothesis of linguistic relativity, the relations between language and perception (and thought) have remained controversial. Recently, anthropological studies of vocabularies of diverse Inuit languages identified over 1,500 terms for sea ice and snow in all Inuit languages; some languages have more than hundred different terms although not all in active everyday use (Krupnik 2011). Even when there exist distinctive cross-cultural patterns of color terms across different languages, it is nonetheless striking how many color terminologies exist across the globe. For human communication, names for colors are unavoidable. In making judgments about color and expressing our thoughts about color, we necessarily use language. Without language, we cannot describe and present perceptual experiences of color. That there are intricate relationships between color and language cannot therefore be denied, but the question is how strong the relation is. A weaker thesis that language influences our perception seems more defensible than the strong thesis that our native language determines how we think and perceive the world. This implies that other factors also play a role, such as our experiences and socio-cultural background (Hussein 2012). We have learned to name, identify and categorize colors with our native vocabulary; it helps us to remember specific colors and to recognize colored objects. Color language is shared with others in the same linguistic community (Maund 1995).

Views regarding the relation between color perception and language are not independent from perspectives on the nature of color, as discussed in the previous chapter. Linguistic relativity seems to presuppose an antirealist theory of colors: they do not actually exist in the outside world but are the product of the mind, in particular language. Perceptual universalism tends towards realist theories assuming that basic color categories can be detected objectively, whether or not they are contingent upon the physical world or the neurophysiology of the perceiver (Saunders 2000). Both assume that colors can be isolated and abstracted from the context in which they are used as well as the community in which color language is evolving and applied. Relationist theories regard color vision as the interaction between perceiver and environment, and as a purposeful activity of the perceiving subject. Humans recognize colors, compare them, and discriminate objects on the basis of color, because color vision has a functional role related to the interest of perceivers (Maund 1995). How colors are named is not simply the detection of qualities of objects nor straightforward construction of the language used. As a property of the perceptual process, color is the result of a visual experience within the life-world of an individual. This world is constituted by history and cultural context in which colors have a particular meaning. Language does have an influence on how we perceive colors, but learning is a social and cultural process: color judgments cannot be regarded as artefacts of language alone.

3.3 The Affective Power of Color

In his *Theory of Colors*, Johann Wolfgang von Goethe (1749–1832) states: "Experience teaches us that particular colours excite particular states of feeling" (Goethe 1970, §762). The book had a wide impact, particularly on artists, color theorists and philosophers. It strongly asserts the idea that colors have an important place in the phenomenal world, and that their effects are directly associated with the emotions of the mind. For Goethe, color is a degree of darkness. Every diminution of light or increment of darkness is increase of color. In his view, color is a primordial phenomenon, located in the mediation of subjective and objective phenomena through turbidity or a semi-transparent medium. This can be observed in the blue of the sky and the yellow of the sun emerging through the cloudy medium of the heavens, which transforms the colors into red depending on atmospheric changes (Goethe 1970, §175, §150). Yellow, blue and red are therefore pure elementary colors.

Fig. 3.1 Joseph Karl Stieler, *Johann Wolfgang von Goethe* (1828). Neue Pinakothek, Munich. Photo by Pierre André (2016), Wikimedia, https://commons.wikimedia.org/wiki/File:Joseph_Karl_Stieler_portrait_de_Johann_Wolfgang_von_Goethe.jpg#/media/File:Joseph_Karl_Stieler_portrait_de_Johann_Wolfgang_von_Goethe.jpg, CC BY-SA 4.0.

From his observations that colors emerge from the contrast between brightness and darkness, Goethe concludes that they present a polarity, what he calls a plus and a minus (Goethe 1970, §696). On the plus side, we have yellow but also action, light, brightness, force, warmth, repulsion; and on the minus side, there is blue, negation, shadow, darkness, weakness, coldness, distance and attraction. When the contrasting sides are combined and in perfect harmony, we call it green. With this polarity, Goethe distinguishes between active and passive colors, but he also revives an older differentiation of warm and cold colors. Some colors (for example orange, red and yellow) evoke images of sunshine and heat. If we see them, warm colors seem to come closer; they make larger rooms cozier. Other colors (such as blue and green) are reminders of sky, water and ice, and they have a calming and soothing effect; they also look as though they recede, making small rooms appear larger. However, the separation of colors as warm or cold seems dependent on the cultural setting. In the medieval period, for example, blue was regarded as a warm color (St Clair 2016). Presumably, the idea that colors could be warm or cold has its origins in the Ancient Greek association between colors and fundamental elements, relating yellow (and red) with fire, and blue with water. The earliest explicit classifications of colors as warm or cold date from the eighteenth century (Gage 1999, 272). In the Romantic era, color systems with a polar contrast between warm and cold became more popular. Studies of the influence of colors on temperature have delivered inconclusive results. In 1926, the first study of the apparent warmth of colors concluded that these colors do not affect one's judgment of the perceived warmth of an object (Morgensen and English 1926). Further studies of the so-called hue-heat hypothesis corroborated the finding that the effect of colors on perceived temperature was small and without practical significance (Fanger, Breum and Jerking 1977; Greene and Bell 1980). More recent studies however confirmed that color has an effect on temperature perception. Ziat and colleagues (2016), for example, found that study participants hold a hot vessel longer when associated with blue, and a cold vessel longer when paired with red. Nonetheless, perceived temperature is not only influenced by color but also by other surrounding factors such as ambient light, humidity and sound. Tsushima and colleagues (2020) found that a sensation of coolness or warmness was manipulated by illumination. In a well-controlled experimental environment participants felt warmer in a room illuminated with warm color, and cold when a cold color illumination was used. Although these results are promising, they do not imply that the hue-heat effect can be practically applied in everyday living situations.

With the notion of polarity of colors, Goethe not only revitalized the contrast between active and passive, and warm and cold colors, but in addition he focused attention on the affective power of colors. They must produce "definite, specific states in the living organ" (Goethe 1970, §761). Goethe discussed the effect of color under the heading of "moral associations," but the examples he provides are more psychological than normative. Colors on the plus side excite feelings that are quick, lively and aspiring. Yellow, for example, produces a warm and agreeable impression. Yellow-red has the highest energy and produces extreme excitement. Colors on the minus side produce anxious and restless impressions. Blue gives a feeling of cold and repose. In the Romantic era, it is a melancholic color; Goethe's Werther is dressed in blue (Pastoureau and Simonnet 2005). Red-blue is associated with an unquiet feeling. The color red leads to an impression of gravity and dignity, and at the same time of grace and attractiveness. Green, as the mixture of yellow and blue, has the effect that "the eye and the mind repose" (Goethe 1970, §802).

Goethe's ideas about the affective power of color have major repercussions in healthcare and psychology. They encourage the development of chromotherapy, i.e. the use of colors in medical practice and treatment, which will be discussed in the next chapter. They also are an important stimulus to color psychology. For the psychiatrist Carl Gustav Jung (1875–1961), the founder of analytical psychology, colors are related to the psyche; they are a reflection of the world of the unconscious, and can therefore reveal psychological traits. Personality types can be linked to the basic colors yellow, red, green and blue (Riley 1995). Jung's ideas are used to assess leadership qualities; for example, red stands for directional leadership (because it reflects a strong-willed and purposeful personality), and blue refers to operational leadership (with a precise and deliberate personality) (Pendleton and Furnham 2016). Another example is the test developed by Max Lüscher (1923–2017) in the 1940s which uses color patches to interpret a subject's character and personality. Blue is associated with passivity, sensitivity, perceptiveness and unity, expressive of tenderness and tranquility. Red, on the other hand, with aggressiveness, offensiveness, autonomy, and competitiveness, expressive of domination, sexuality and desire (Gage 1999, 32). The assumption is that color preferences are the result of meanings associated with colors, and that they thus reveal who people are and what they value. Whereas the test is widely used in personality testing, for example in psychology, management and personnel selection, and many claims are made in popular media that character and personality can be revealed by color preferences, the reliability and reproducibility of the Lüscher color test are contested. For example, Lüscher's prediction that people favoring red and yellow are more extraverted than those who prefer blue and green could not be confirmed (Picco and Dzindolet 1994). Systematic analyses of popular media claims discovered no reliable associations between color preferences and personality traits (Jonauskaite et al. 2021).

3.4 The Meaning of Color

That color is the language of emotional affect is expressed by numerous artists. William Turner (1775–1851) was inspired by Goethe and two of his paintings directly refer to the German poet (König and Collins 2009).

Fig. 3.2 William Turner, *Light and Colour (Goethe's Theory)* (1843). Tate Britain, London. Photo by Wuselig (2020), Wikimedia, https://commons.wikimedia.org/wiki/File:Horror_und_Delight-Turner-Light_and_Colour_(Goethe%27s_Theory)_DSC2252.jpg#/media/File:Horror_und_Delight-Turner-Light_and_Colour_(Goethe's_Theory)_DSC2252.jpg, CC0 1.0.

For him, color has first of all a symbolic value; it is not simply a sensation of darkness and light (Gage 1999, 165). Kazimir Malevich (1879–1935) advocated art as the expression of human feelings, rather than as the reflection of the visual world or an imitation of nature. Feeling is the only source of creation: "An artist who creates rather than imitates expresses himself; his works are not reflections of nature but, instead, new realities, which are no less significant than the realities of nature itself" (Malevich 1959, 30). The visual phenomena of the objective world are, in themselves, meaningless; feeling is the only significant thing. The supremacy of pure feeling leads to abstract, non-objective art which evokes meaning through shapes and forms of color.

Fig. 3.3 Utigawa Kuniyoshi, *Medical and Surgical Treatments for a Lame Princess and Others* (1849/52). Wellcome Collection. Wikimedia, https://commons.wikimedia.org/wiki/File:Medical_and_surgical_treatments_for_a_lame_princess_Wellcome_L0035015.jpg#/media/File:Medical_and_surgical_treatments_for_a_lame_princess_Wellcome_L0035015.jpg, CC BY 4.0.

The idea that colors have psychological effects is related to the belief that they have a meaning or are associated with particular meanings. In history and across cultures specific colors are connected to particular meanings. For example, red is the archetypical color; the first color to be used in painting (e.g. in the prehistoric paintings in the Cave of Altamira) and dyeing. In many languages, the same word refers to 'red' and 'colored' (Pastoureau 2017). The primary role of red is usually explained because this color refers to blood and fire as two natural elements and experiences of all societies; it is considered the color of life. Already in ancient civilizations, red is associated with power and the sacred. In Japanese culture, red is an important color, representing strong emotions. It is used as a symbol of the innocence of youth, especially in young women and during wedding ceremonies. It is also associated with authority and wealth, while specific shades of red are protective against evil and disaster.

The historical picture shows a princess, daughter to the Emperor of Japan, who was physically impaired and did not want people to know it. She is shown in the right panel, dressed in red, being attended by a doctor or surgeon in black. All figures in black are doctors or surgeons. In Roman traditions, red is the color of strength, energy and victory. It remains associated with power in the medieval West, particularly as the symbolic color of justice and as the dominant color in coats of arms. It also becomes the color of love, the expression of affection and passion.

However, red began to lose its primary status in the fourteenth century. In Western civilization, new color codes emerged that consider red as an indecent and immoral color. Negative connotations come to the foreground through red's evocation of the flames of hell, and thus violence, sin and crimes. It becomes the color of executioners, traitors and rebels. Newton's discoveries remove red from the center of the chromatic scale (between white and yellow on the one hand, and green, blue and black on the other) to one end of the visible light spectrum, though it continues to be regarded as a primary color. Red turns into a warning signal of imminent danger and a sign of prohibition, introduced in international signaling systems. As a vital power, red attracts attention and is therefore the color of fire engines, humanitarian organizations (such as Red Cross and Red Crescent), and red-light districts (Pastoureau 2017).

Similar duplicities of meaning are noticeable for other colors. While green, for example, nowadays refers to nature and cleanliness, in most European societies it has long been regarded as an intermediary, non-violent, soothing and peaceful color (between white, red and black), playing a minor role in social life and artistic imagination (Pastoureau 2014). In Ancient Egypt, green is regarded as a beneficial color, meaning fertility, growth, youth and regeneration. In Islamic civilizations, green is always a positive color associated with spring, the sky and paradise, a symbol of happiness and hope. In the West, green has, for a long time, been treated as a dangerous and chemically unstable color. It refers to everything that changes, varies and moves, and is thus interpreted as the color of chance, games, money, hope and fate. In medieval society, it was the color reserved for ordinary people, i.e. servants and peasants. It also carried an element of ambiguity, as the pigments used to create green paint were often poisonous. Demons, bad spirits, dragons and other frightening creatures are often presented in green. For a long time, green was not considered as a primary color but the product of mixing blue and yellow. As a secondary color, it was not commonly used, while blue became the preferred color, particularly during the Enlightenment in Europe. The prism experiments of Newton introduced a new color order with green placed between blue and yellow. In later color theories, it is considered as opposite to red, and becomes the color of permissiveness. This is first observed in the code for maritime signals, with green giving ships access to the port, and in the nineteenth century in city traffic and railroads. Today, receiving the 'green light' has taken on a broader meaning, symbolizing permission to go ahead. Since the Romantic era, appreciation for green has increased. The color has become closely associated with nature, ecology and hygiene, as a symbol of freedom, youth and health. Regarded as a source of renewal and a sign of social responsibility and ecological sustainability, the color green now reflects a political and ethical stance in many societies (Pastoureau 2014).

Blue is now the favorite color among people in Europe and North America. However, for much of history, it was neither common nor highly valued. In classical Antiquity, blue was not even considered a color, and in Greek texts words for blue are absent. In ancient Rome, blue is the color of barbarians and strangers (Pastoureau and Simonnet 2005). Words for blue are not derived from Latin but from Germanic and Arabic origins. The meaning of blue starts to change in Western civilizations in the twelfth and thirteenth centuries (Pastoureau 2001). Religious discourse, especially the theology of light, imagines God as light, and blue becomes a divine color. In the Cult of the Virgin, the coat and dress of Mary is blue. Many churches install blue stained windows. In turn, the system of basic colors changes from three (white, red and black) to six (plus blue, green and yellow), posing blue as the opposite of red. A new color order starts to take shape with blue as the fashion color of the aristocracy. At the end of the Middle Ages, a moralistic interpretation of colors includes blue (with white, black, grey and brown) in the category of worthy colors, viewed as sober, discrete and dignified. New techniques of producing blue pigments (e.g. Prussian blue and the massive import of indigo) make blue highly fashionable in Western societies. Blue is a color that is not shocking but calm, wise, modest, peaceful and timid (Pastoureau 2001).

That colors have particular positive and negative meanings is an important issue in social and historical studies, as illustrated in the examples discussed above. They demonstrate at least three things. First, preferences for color may vary; they have ups and downs; one color may be the favorite at one time, due to technical matters (for example, the availability of pigments), economic considerations (some pigments may be expensive and are therefore primarily used by the rich and powerful), religious and political factors determining by whom and under what circumstances they may be used, and cultural trends reflected in fashion. Second, the meaning of color changes with time and cultural setting. Most studies of color history have focused on Western cultures. It cannot be assumed that similar preferences prevail in other cultures. While blue is now the color of choice in the West, white is the most favored color in Asian countries, associated with cleanliness, purity, harmony, chastity and light (Saito 1996). Yellow has long been an unpopular color in Europe, associated with negative connotations such as envy, jealousy, dishonor and treason, while in China it used to be a sacred, imperial color, and in the Buddhist tradition refers to humility and renunciation (Pastoureau 2019).

Fig. 3.4 Yellow emperor. Scan from *Shèhuì Lìshǐ Bówùguǎn* [Social History Museum]. Wikimedia, https://commons.wikimedia.org/wiki/File:Yellow_Emperor.jpg#/media/File:Yellow_Emperor.jpg, public domain.

Third, the meaning of color depends on the context. Whereas blue is commonly the most preferred color, it is not appreciated in the context of food. The meaning of colors can differ when they are used for dress, coats of arms, liturgy, traffic or interior design. In Christian culture, green is the liturgical color for ordinary Sundays because it is a middle color between white, red and black. As a neutral color it is not regarded as dignified and active, and thus for a long time it played a minor role in daily life. According to Pastoureau (2014, 206), green was rarely used for everyday objects, décor and furniture until the 1950s. At the same time, it has always been the color of hope. Greenland was named as such not because of its abundant vegetation but as a way to inspire hope and attract settlers with the promise of a hospitable land. Associated with spring, green also referred to new life and youth, and was the typical color for pregnant women. Since medication used to be derived from herbs and plants, green has been the emblematic color of medicine and pharmacology; in many countries pharmacies still use a green cross for recognition. Nowadays, green is popular as the color of ecology, a mark of environmental responsibility and sustainability. This association between meaning and context is coherent with relationist theories of color. Perception of color is not merely a matter of observation and identification but is functional: it helps to engage with the world, dependent on the interests of the perceiver. The interaction of meaning and context or use is dialectical. Certain colors are chosen for objects and activities because of their connotations (meaning influences use). At the same time, our familiarity with colored objects teaches us the meanings of those colors (use determines meaning). This implies that colors do not have a meaning inherent to themselves, prior to perception and independent of the cultural and social contexts in which they are perceived. Instead, humans have become accustomed to colors as meaningful in a specific context, and have learned, mostly implicitly, the connotation and significance of colors in practices of daily life.

It does not follow that the meanings of color are determined by society and culture alone. In all civilizations and cultures humans share similar experiences: green refers to vegetation, growth and spring, blue to water and sky, and red to blood and heat. It is not surprising that studies repeatedly demonstrate significant agreements in the connotations of colors. Students from twenty countries in different continents consistently evaluated black as bad, strong and passive, red as strong and active, yellow as weak, blue and green as good. The authors suggest that there are strong universal trends in the attribution of affective meaning to colors, even though their data show the ubiquity of exceptions to general trends (Adams and Osgood 1973). While this study tested responses to color concepts, another study with students in the United States examined colored words. It confirmed that red is associated with failure and green with success (Moller et al. 2009). When actual swatches of color are used to identify their meanings, respondents from eight different countries all regarded blue, green and white as "peaceful," "gentle" and "calming" while red is associated with "active," "hot" and "vibrant" (Madden et al. 2000). In another study, when participants are not asked to respond to actual color samples but to imagine their own examples of target colors, similar meaning associations are found: red is perceived as a warm color, exciting and stimulating, while blue and green are calm, soothing and peaceful; black is bad and regarded as a symbol of evil, malice and death; pink is associated with femininity; brown and grey are weak or negative, lacking emotional quality (Clarke and Costall 2008).

The conclusion from these studies is that colors seem to be associated with meanings when subjects are confronted with concepts, words, actual and imagined colors. It is therefore not clear that the perception of color itself generates meanings. Research on color is criticized because of methodological weaknesses. First, color stimuli are often not specified: is the investigation directed on the effects of hue or of saturation and brightness? When color samples are used, they are sometimes not related to a standardized system of color notation (such as the Munsell System, specifying hue, saturation and brightness). Also, the lighting conditions in the research are frequently unspecified and not standardized. Second, the concept of meaning is not defined; if subjects are asked to attribute meanings to color it is not evident that they share similar connotations when discrete meanings (such as "exciting" or "soothing") are mentioned (Valdez and Mehrabian 1994).

3.5 Color, Emotions and Feelings

Goethe's theory that colors have an immediate effect on the emotions of the mind has inspired numerous studies of the impact of colors on emotions and feelings. College students in the United States, for example, respond most positively to green and yellow (using color samples from the Munsell System). Green elicits relaxation and calmness, but also happiness, peace and hope, creating feelings of comfort. Yellow is regarded as lively and energetic, with its associations with the sun, blooming flowers and summertime. White is viewed positively, referring to feelings of innocence, peace and hope, purity and simplicity. It can also have negative responses of emptiness, loneliness and boredom. Black evokes negative emotions such as sadness, depression, fear and anger, associated with death, mourning and tragic events while richness, wealth and power are identified as positive aspects (Kaya and Epps 2004). A study in five different countries (Germany, Mexico, Poland, Russia and the USA) using color terms observed that in all nations anger is associated with black and red, and jealousy with red (Hupka et al. 1997). However, a study that uses color samples from the Munsell System, with standardized background and lighting conditions, and experimental controls to investigate the effects of hue, saturation and brightness, provides consistent evidence that emotional reactions are strongly related to color brightness and saturation, rather than hue (Valdez and Mehrabian 1994).

Regardless of the evidence, the connection of colors and emotions is often mentioned in the literature, especially popular media. The ancient distinction between warm and cold colors is frequently repeated, now explained in scientific terms with references to physics and physiology. Work on the relation between color psychology and physiology was inspired by the experiments of the German neurologist and psychiatrist Kurt Goldstein (1878–1965) on the effects of color on the human body. Under the influence of Goethe's theory, Goldstein argued that colors influence the physiology of the human organism, observable, for example, in emotional responses, behavior, the position of the body and motor action. He also argued that specific colors produce specific response patterns of the organism. Red stimulates activity, and creates the emotional background in which ideas and actions will emerge, whereas green generates the conditions to develop ideas and executive actions (Goldstein 1942). Goldstein's ideas are later interpreted in terms of wavelength and arousal (Elliot and Maier 2012). Colors at one end of the electromagnetic spectrum (warm ones such as red and yellow) are long wavelengths that are more arousing and stimulating than the short wavelength colors at the other end of the spectrum (cold ones like blue and green). The argument is that experimental studies of physiological measures have corroborated these effects. This has led Birren to conclude that in human beings, red "tends to raise blood pressure, pulse rate, respiration, and skin response (perspiration) and to excite brain waves... Blue tends to have reverse effects, to lower blood pressure and pulse rate. Skin response is less, and brain waves tend to decline. The green region of the spectrum is more or less neutral" (Birren 1978, 24). Critical examination of the available evidence confirms that there are indeed nonvisual physiological responses to color. Red has an arousal effect when measured by electroencephalograms and galvanic skin response. The data on blood pressure, respiration and heart rate are inconclusive. The problem is that it is difficult to decide whether the effects of color are directly the result of visual stimulation, or mediated indirectly by cognitive associations. In other words, if arousal is measurable when people see red, is this a physiological response, the effect of associated meanings, or both? (Kaiser 1984). Recent research with colors controlled for saturation and with controlled lighting conditions could not detect that red is more arousing than blue (Fehrman and Fehrman 2018, 88). If there is a difference in response to colors, it is not the result of physiology but of cultural and learning processes that associate specific colors with particular meanings. The researchers conclude: "Colors do not inherently contain excitement or calming effects" (Fehrman and Fehrman 2018, 91). This conclusion does not entail that colors are inept; on the contrary, they have vast effects on human beings but these are not due to the colors themselves but to the meanings that are associated with them, and furthermore due to the fact that color and light are inseparable. As will be discussed in the next chapter, light has diverse effects on biological systems and influences chemical reactions in the organism. It is almost impossible to separate the effects of color and light.

3.6 Color, Human Behavior and Performance

If colors are connected with meanings and emotions, this may have implications for psychological well-being and functioning. A famous example is a study reporting that the color pink has a calming effect and reduces aggression (Schauss 1979). This finding was immediately used in some correctional facilities in the United States: holding cells for initial confinement of new inmates were painted pink, whereupon no incidents of violent and hostile behavior occurred (Fehrman and Fehrman 2018) However, the effects of the pink room were only temporary, and subsequent studies could not confirm them (Elliott and Maier 2012).

The possible influence of colors on human behavior and performance is the subject of a range of studies in various settings. One finding is that people make more proofreading errors in white interior offices than in blue and red offices, even if they mostly prefer to work in white and beige offices since these are regarded as the least distractive colors (Kwallek et al. 1996). Another finding is that the learning performance of students is best when the walls in their study rooms are blue. Red-colored walls have a negative impact on intellectual activity since they impair concentration (Al-Ayash et al. 2016). In contexts like these in which an achievement is expected, perception of red diminishes performance, particularly when cognitive analysis, mental manipulations and flexible processes are required (Maier, Elliott and Lichtenfeld 2008). On the other hand, blue enhances performance of a creative task. If creativity and imagination are required (for example, in the development of a new product or in a brainstorming session) blue is more beneficial than red (Mehta and Zhu 2009).

That the context is important for the effects of colors is shown in studies about contests. In a competitive context, the color red influences the outcome of a contest. Wearing a red outfit is associated with a higher probability of winning. In the 2004 Olympic Games, contestants in combat sports (e.g. boxing, wrestling and tae kwon do) wearing red significantly won more fights than those in blue. In the Euro 2004 international soccer tournament those with red shirts scored more goals than those in other colored shirts (Hill and Barton 2005). Furthermore, in relational contexts, red is a positive color. Red clothing enhances the attractiveness of women. This is only an effect on men, not on women themselves, while men are unaware of this effect (Elliott and Niesta 2008).

The idea that the influence of color on performance and psychological functioning is dependent on the context is elaborated in the color-in-context theory of Elliot and Maier (2012). The assumption is that colors not only have aesthetic value but also carry meaning which generates processes of evaluation and appraisal that influences affect, cognition and behavior. These processes occur automatically without the perceiver being aware of them. The specific meaning of colors according to this theory is the result of classical conditioning (we have learned them since infancy) and societal learning (the significance of colors is experienced and reiterated within social and cultural life). Some color associations may be rooted in biology since specific colors are relevant for the adaptation and survival of the organism. The thesis of the theory is that any color can have diverse meanings but that its distinct meaning is context dependent. Colors are not perceived as isolated and abstract phenomena but always connected to other visual qualities of the surrounding world. Red refers to edibility and pleasure when viewed in strawberries, but its meaning changes to a warning signal if we participate in traffic. Similarly, a colored object may have a different meaning in a different psychological context, depending on what we intend to do and how we are emotionally affected. The theoretical assumptions of the color-in-context theory are much in line with relationist theories of the nature of color, that regard color as a relational property, as interaction between object, environment and perceiver, and that interpret color vision as an activity with various functions (see Chapter 2). Elliot and Maier (2012) discuss the empirical research that seems to confirm their theory. In achievement contexts, in which competence is evaluated, red impairs intellectual performance since it has the meaning of danger or failure. It therefore evokes avoidance motivation in such contexts (Elliot et al 2007). On the other hand, in affiliation contexts involving heterosexual interaction, red has a positive meaning: it facilitates the attraction of males to females and the attraction of females to males. Although most studies have focused on the influence of red on psychological functions, connections have also been described for other colors. Experiments exposing participants to hues of equal saturation and lightness show that green—with its connotations of growth, development and success—has a positive influence on creative performance (more than white, grey, blue and red) (Lichtenfeld et al. 2012). Blue has also a positive impact on creativity when compared to red (Mehta and Zhu 2009).

3.7 Practical Implications

If perception of color depending on the context has an impact on affect, cognition and behavior of people, it is obvious that there are practical implications. A famous example is waitresses who are wearing red receiving more frequent and higher tips from male (not female) customers, than waitresses wearing other colors (Guéguen and Jacob 2014): a finding suggesting how waitresses could increase their income. Much empirical work on color and psychological functioning has focused on applied questions and the use of color for particular purposes, often without theoretical framework.

One area of study is marketing. When brands have a particular color, it helps us recognize the brand and further establishes a visual identity which communicates a certain image, producing a distinctive personality for the brand. Red associates a brand with excitement, white with sincerity, and blue with competence (Labrecque and Milne 2012). Decisions to purchase a product are not only based on price and quality but also on color. It is important to know what meanings consumers ascribe to the color of products. A cross-cultural study in eight countries in different continents indicates that all subjects prefer blue, green and white for product logos since these colors are associated with "peaceful," "gentle" and "calming" effects, while red has "hot" and "vibrant" associations across all countries. For other colors, substantial differences in meaning are found. Since colors may invoke different consumer reactions, marketing managers should examine how colors are perceived in specific countries and cultures (Madden, Hewett and Roth 2000).

Another area of study is interior design (Enwin et al 2023). It is evident that aesthetics here play a major role. Colors should express the taste of consumers and should be pleasing. At the same time, numerous recommendations and instructions abound on how to use colors in interior spaces, appealing to their psychological effects. Since people spend much time within residential spaces it is important to understand the impact of interior colors on mood and well-being. It is crucial how colors are combined; in interior spaces they should be in harmony. The effect of colors is furthermore dependent on the social function of an interior; it is different for theatres, hospitals, offices, museums and homes. Within homes, the function of a space determines which colors are most appropriate. Recommendations are most often based on the distinction between warm and cold colors. Since red, for example, stimulates excitement, activity and appetite it can be used in the design of kitchens and dining rooms. Cold colors such as blue, on the other hand, are associated with calmness and relaxation, and are most suitable for bedrooms and study spaces (Ćurčić et al. 2019). Likewise, colors are used to enhance the spatial experience of interiors (Cao 2019). Warm colors seem to advance to us, bringing objects to the foreground, whereas cold colors seem to have a receding effect. Blue is therefore used to make small rooms seem larger while advancing colors like red give the illusion that large rooms are smaller. Recommendations about the application of colors in interior design are difficult to interpret. Most studies are done in laboratory settings. In real situations, many factors besides color play a role, such as lighting and materials, as well as preferences, cultural setting and fashion trends (Fehrman and Fehrman 2018).

The effects of colors are extensively studied in relation to food. The quality of meat, fish and fruit is immediately appraised by their color. This is often the most powerful visual aspect of food packaging, influencing consumer decisions. Red packaging is associated with hot flavors, and green with nature and environmental friendliness (Yu et al. 2021). Another question is whether the color of food influences taste and the perception of flavor. Studies document that red food coloring has an effect on the perception of sweetness (in nature, ripe fruits are red and thus sweet). Color has no influence on the perception of saltiness (Spence et al. 2010). An interesting topic is the relationship between color and food consumption. A finding that has drawn substantial popular attention, especially among those trying to lose weight, is that people eat less when food is served on red plates. Researchers served participants popcorn and chocolate chips on either red, white or blue plates. Those who ate from the red plates, consumed less than the others

(Bruno et al. 2013). The same is true for snacks and soft drinks: red labels significantly reduce their consumption (Genschow et al. 2012). It is unclear how to explain the color red's effect on consumption, though it is assumed that this color signals danger and prohibition, so that at a non-conscious level it induces avoidance behavior. This is a significant finding for weight-loss programs and for the treatment of eating disorders (Bruno et al. 2013); it is also a helpful approach for reducing the intake of unhealthy food and drink (Genschow et al. 2012) in general, because it does not require medical interventions or medications. The Dutch National Institute for Public Health and the Environment advocates for the use of color in discouraging consumption of harmful substances, specifically in relation to smoking: instead of white (the color required by law), cigarettes should have a darker color, such as green (associated with harm and damage to health) (RIVM 2023). The meaning of green here is not interpreted as permissive but as poisonous.

3.8 Conclusion

Colors are everywhere around us and they are not inert. They affect us and we cannot remain indifferent to them. This chapter has discussed the influences of colors on human emotions and behavior. Since chromatic experiences are touching and meaningful, color is characterized as a language without words; it immediately and unconsciously communicates emotion and significance. The affective power of colors can be experienced in multiple ways, through viewing natural scenery, looking at paintings in a museum, or designing a website with colorized software. The colors perceived are named, classified and interpreted with words available in our language, and we have learned to identify and distinguish them in social learning processes. Specific languages vary in the number of names they employ to designate colors, but all languages have a significantly smaller lexicon of color names than the total range of colors that humans are, in principle, able to discriminate. It is controversial how the relation exactly works between language and color perception. Is color vision dependent on the linguistic system of the perceiver so that a speaker of a language that does not have words for a particular color does not see that color? The discovery that different languages have a restricted set of similar color terms has made the thesis of linguistic relativity less plausible. Perceiving is an activity that distinguishes a limited number of categories of colors, not as abstract or isolated qualities but always connected to objects or entities, and embedded within specific contexts. This perceptual experience is reflected in the finding that various languages share the same basic color terms: they have minimally two, and maximally eleven, basic terms.

Since the studies of Goethe in the Romantic era, the affective power of colors has attracted growing attention. Characterizing colors as active (plus) or passive (minus), Goethe revived a much older tradition of presenting colors as warm or cold. They evoke a range of particular feelings between excitement and energy, on the one hand, and relaxation on the other. Goethe relates this to the associations that colors generate; although he calls these associations moral, they are in fact in the first place psychological, since they have effects on feelings and emotions. These ideas encouraged subsequent developments, for example the use of colors in medical therapy, and in psychology. Colors are used to make psychological profiles and to test personality traits, often in a simplified way.

Relating the power of colors to their generated associations directs attention to the notion of meaning. Historical studies show that attributing particular meanings to colors is very common in a wide variety of cultures. Most colors have a positive as well as negative meaning depending on the context, availability of pigments, cultural setting and historical period. Although these meanings have changed over time, and some colors today have significantly different connotations than in the past, there is also a remarkable consistency over time, with ancient associations persisting today. However, when colors are intrinsically and unconsciously connected to specific meanings, the question remains as to how we can explain their effects. Are they the result of color alone or are more factors at play? The notion of meaning articulates the importance of language, but it also emphasizes the special significance of things and events perceived or the value that they have for someone. It is a notion that highlights the interaction between perceived object, perceiver and context. Relationist theories of colors are most pertinent to explain this. Colors do not have a meaning in themselves before they are perceived, and separated from the context in which they are seen. We have learned the significance of colors in practices of daily life and in communicating with others in society and culture. This conclusion is relevant for our later analysis of the role of color in ethical debates (Chapter 6). If colors are powerful because they are associated with meanings, and generate normative intuitions of goodness or badness, as well as aesthetic affections of beauty or ugliness, and if these meanings and associations are not generated by colors themselves, it is imperative to focus on the social and cultural processes and practices in which the significance of colors is learned, transmitted and affirmed.

As discussed in this chapter, research on colors has confirmed that they have effects on emotions and feelings, as well as on behavior and performance. At the same time, it is difficult to conclude that the effects are produced by the colors themselves, the meanings associated with them, or both. The difficulty is that many studies are not rigorous and controlled for various factors. First, colors have the dimensions of hue, saturation and lightness, and research rarely distinguishes between these dimensions. Second, it is clear that light has an influence on the human organism, but frequently no distinction is or can be made between hue and light. Third, many studies are performed in experimental, laboratory conditions in which the context of color vision is disregarded and only the effect of colors are tested. For these reasons, it is often not clear that the experimental findings can be applied in real, everyday scenarios. This does not imply that perception of color has no effect. What is needed is a theoretical framework (such as the color-in-context theory mentioned above) to explain the effects of particular colors.

Regardless of how their effects can be theoretically explained, investigations of colors frequently are practically motivated. It is relevant to know which colors make products most attractive to consumers, or what colors influence the mood of people in interior spaces. Particularly in the areas of marketing, interior design and food production and packaging, numerous studies of the impact of colors have been carried out. The underlying assumption of these studies is that the presence or perception of particular colors make people more comfortable, enhancing their mood and well-being, while other colors are unpleasant and disagreeable, initiating avoidance behavior. This assumption also attributes an important role to colors in the context of healthcare, the subject of the next chapter.

References

Adams, F. M. and Osgood, C. E. 1973. A cross-cultural study of the affective meanings of color. *Journal of Cross-Cultural Psychology* 4 (2): 135–156.

Al-Ayash, A., Kane, R. T., Smith, D., and Green-Armytage, P. 2016. The influence of color on student emotion, heart rate, and performance in learning environments. *Color Research and Application* 41: 196–205, https://doi.org/10.1002/col.21949

Berlin, B. and Kay, P. 1991. *Basic color terms. Their universality and evolution.* Berkeley, CA: University of California Press (original 1960).

Birren, F. 1978. *Color & human response. Aspects of light and color bearing on the reactions of living things and the welfare of human beings.* New York: Van Nostrand Reinhold Company.

Bruno, N., Martani, M. Corsini, C., and Oleari, C. 2013. The effect of the color red on consuming food does not depend on achromatic (Michelson) contrast and extends to rubbing cream on the skin. *Appetite* 71: 307–313, https://doi.org/10.1016/j.appet.2013.08.012

Cao, Y. 2019. Analysis of the uses of color elements in architectural interior design. *Journal of World Architecture* 2 (2): 9–13, https://doi.org/10.26689/jwa.v2i2.591

Clarke, T. and Costall, A. 2008. The emotional connotations of color: A qualitative investigation. *Color Research and Application* 33 (5): 406–410, https://doi.org/10.1002/col.20435

Ćurčić, A. A., Keković, A., Ranđelović, D., and Momčilović-Petronijević, A. 2019. *Effects of color in interior design.* 7th International Conference Contemporary Achievements in Civil Engineering. April, Subotica, Serbia, http://zbornik.gf.uns.ac.rs/doc/NS2019.080.pdf

Elliot, A. J., Maier, M. A., Boller, A. C., Friedman, R., and Meinhardt, J. 2007. Color and psychological functioning: The effect of red on performance attainment. *Journal of Experimental Psychology. General* 136 (1): 134–168, https://doi.org/10.1037/0096-3445.136.1.154

Elliott, A. J., and Niesta, D. 2008. Romantic red: Red enhances men's attraction to women. *Journal of Personality and Social Psychology* 95 (5): 1150–1164, https://doi.org/10.1037/0022-3514.95.5.1150

Elliott, A. J., and Maier, M. A. 2012. Color-in-context theory. *Advances in Experimental Social Psychology* 45: 61–125, https://doi.org/10.1016/b978-0-12-394286-9.00002-0

Enwin, A. D., Ikiriko, T. D., and Jonathan-Ihua, G. O. 2023. The role of colours in interior design of liveable spaces. *European Journal of Theoretical and Applied Sciences* 1 (4): 242–262, https://doi.org/10.59324/ejtas.2023.1(4).25

Fanger, P. O., Breum, N. O., and Jerking, E. 1977. Can colour and noise influence man's thermal comfort? *Ergonomics* 20 (1): 11–18, https://doi.org/10.1080/00140137708931596

Fehrman, C., and Fehrman, K. 2018. *Color. The secret influence*. Solana Beach, CA: Cognella Academic Publishing (4th edition).

Gage, J. 1999. *Colour and meaning. Art, science and symbolism*. London: Thames & Hudson.

Genschow, O., Reutner, L., and Wänke, M. 2012. The color red reduces snack food and soft drink intake. *Appetite* 58: 699–702, https://doi.org/10.1016/j.appet.2011.12.023

Goethe, J. W. von. 1970. *The theory of colours*. Translated by C. L. Eastlake. Cambridge, MA: MIT Press (original 1810).

Goldstein, K. 1942. Some experimental observations concerning the influence of colors on the function of the organism. *Occupational Therapy* 21: 147–151.

Greene, T. C., and Bell, P. A. 1980. Additional considerations concerning the effects of 'warm' and 'cool' wall colours on energy conservation. *Ergonomics* 23 (10): 949–954.

Guéguen, N., and Jacob, C. 2014. Clothing color and tipping: Gentlemen patrons give more tips to waitresses with red clothes. *Journal of Hospitality & Tourism Research* 38 (2): 275–280, https://doi.org/10.1177/1096348012442546

Hardin, C. L. 1993. *Color for philosophers. Unweaving the rainbow*. Indianapolis, IN: Hacket Publishing Company.

Hill, R., and Barton, R. 2005. Red enhances human performance in contests. *Nature* 435: 293.

Hupka, R. B., Zaleski, Z. Otto, J., Reidl, L., and Tarabrina, N. V. 1997. The colors of anger, envy, fear, and jealousy. A cross-cultural study. *Journal of Cross-cultural Psychology* 28 (2): 156–171.

Hussein, B. A. 2012. The Sapir-Whorf hypothesis today. *Theory and Practice in Language Studies* 2 (3): 642–646, https://doi.org/10.4304/tpls.2.3.642-646

Jonauskaite, D., Thalmayer, A., Müller, L., and Mohr, C. 2021. What does your favourite colour say about your personality? Not much. *Personality Science* 2: e6297, https://doi.org/10.5964/ps.6297

Kaiser, P. K. 1984. Physiological response to color: A critical review. *Color Research and Application* 9 (1): 29–36.

Kay, P. 2005. Color categories are not arbitrary. *Cross-Cultural Research* 39 (1): 39–55, https://doi.org/10.1177/1069397104267889

Kaya, N., and Epps, H. 2004. Relationship between color and emotion: A study of college students. *College Student Journal* 38 (3): 396–405.

König, C. S., and Collins, M. W. 2009. Goethe, Eastlake and Turner: From colour theory to art. *International Journal of Design & Nature and Ecodynamics* 4 (3): 228–237.

Krupnik, I. 2011.'How many Eskimo words for ice?' Collecting Inuit sea ice terminologies in the International Polar Year 2007–2008. *The Canadian Geographer/Le Géographe canadien* 55 (1): 56–68, https://doi.org/10.1111/j.1541-0064.2010.00345.x

Kwallek, N., Lewis, C. M., Lin-Hsiao, J. W. D., and Woodson, H. 1996. Effects of nine monochromatic office interior colors on clerical tasks and worker mood. *Color Research and Application* 21 (6): 448–458.

Labrecque, L. I., and Milne, G. R. 2012. Exciting red and competent blue: The importance of color in marketing. *Journal of the Academy of Marketing Science* 40: 711–727, https://doi.org/10.1007/s11747-010-0245-y

Lichtenfeld, S., Elliot, A. J., Maier, M. A., and Pekrun, R. 2012. Fertile green: Green facilitates creative performance. *Personality and Social Psychology Bulletin* 38 (6): 784–797, https://doi.org/10.1177/0146167212436611

Madden, T. J., Hewett, K., and Roth, M. S. 2000. Managing images in different cultures: A cross-national study of color meanings and preferences. *Journal of International Marketing* 8 (4): 90–107, https://doi.org/10.1509/jimk.8.4.90.19795

Maier, M. A., Elliott, A. J., and Lichtenfeld, S. 2008. Mediation of the negative effect of red on intellectual performance. *Personality and Social Psychology Bulletin* 34 (11): 1530–1540, https://doi.org/10.1177/0146167208323104

Malevich. K. 1959. *The non-objective world*. Chicago, IL: Paul Theobald and Company.

Maund, B. 1995. *Colours. Their nature and representation*. Cambridge, UK: Cambridge University Press.

Mehta, R., and Zhu, R. 2009. Blue or red? Exploring the effect of color on cognitive task performances. *Science* 323 (5918): 1226–1229, https://doi.org/10.1126/science.1169144

Mogensen, M. F., and English, H. B. 1926. The apparent warmth of colors. *The American Journal of Psychology* 37 (3): 427–428, https://doi.org/10.2307/1413633

Moller, A. C., Elliot, A. J., and Maier, M. A. 2009. Basic hue-meaning associations. *Emotion* 9 (6): 898–902, https://doi.org/10.1037/a0017811

Pastoureau, M. 2001. *Blue. The history of a color*. Princeton, NJ and Oxford: Princeton University Press.

Pastoureau, M. 2014. *Green. The history of a color*. Princeton, NJ and Oxford: Princeton University Press.

Pastoureau, M. 2017. *Red. The history of a color*. Princeton, NJ and Oxford: Princeton University Press.

Pastoureau, M. 2019. *Yellow. The history of a color*. Princeton, NJ and Oxford: Princeton University Press.

Pastoureau, M., and Simonnet, D. 2005. *Le petit livre des couleurs*. [The little book of colors] Paris: Éditions du Panama.

Pendleton, D., and Furnham, A. 2016. *Leadership: All you need to know*. London: Palgrave/Macmillan, https://doi.org/10.1057/978-1-137-55436-9

Picco, R. D., and Dzindolet, M. T. 1994. Examining the Lüscher Color Test. *Perceptual and Motor Skills* 79 (3): 1555–1558.

Pullum, G. K. 1991. *The great Eskimo vocabulary hoax and other irreverent essays on the study of language*. Chicago, IL and London: The University of Chicago Press.

Riley, C. A. 1995. *Color codes. Modern theories of color in philosophy, painting and architecture, literature, music, and psychology*. Hanover and London: University Press of New England.

RIVM (Dutch Institute for Public Health and the Environment). 2023. Darker cigarettes and other measures to make cigarettes less appealing. *RIVM*, 14 April, https://www.rivm.nl/en/news/darker-cigarettes-and-other-measures-to-make-cigarettes-less-appealing

Romano, C. 2020. *De la couleur*. [About color] Paris: Éditions Gallimard.

Saito, M. 1996. A comparative study of color preferences in Japan, China and Indonesia with emphasis on the preference for white. *Perceptual and Motor Skills* 83: 115–128.

Saunders, B. 2000. Revisiting Basic color terms. *Journal of the Royal Anthropological Institute* 6 (1): 81–99.

Schauss, A. G. 1979. Tranquilizing effect of color reduces aggressive behaviour and potential violence. *Journal of Orthomolecular Psychiatry* 8: 218–220.

Spence, C., Levitan, C. A., Shankar, M. U., and Zampini, M. 2010. Does food color influence taste and flavor perception in humans? *Chemosensory Perception* 3 (1): 68–84, https://doi.org/10.1007/s12078-010-9067-z

St Clair, K. 2016. *The secret lives of color*. London: Penguin.

Street, B. 2018. *Art unfolded: A history of art in four colours*. Lewes: Ilex Press.

Tan, J., van Onna, H., and Kamphuis, H. (eds). 2011. *Colour hunting. How colour influences what we buy, make and feel*. Amsterdam: Frame Publishers.

Thompson. E. 1995. *Colour vision. A study in cognitive science and the philosophy of perception*. London and New York: Routledge.

Tsushima, Y., Okada, S., Kawai, Y., Sumita, A., Ando, H., and Miki, M. 2020. Effect of illumination on perceived temperature. *PLoS ONE* 15 (8): e0236321, https://doi.org/10.1371/journal. pone.0236321

Valdez, P., and Mehrabian, A. 1994. Effects of color on emotions. *Journal of Experimental Psychology: General* 123 (4): 394–409.

Whorf, B. L. 1940. Science and linguistics. *Technology Review* 42 (6): 1–6.

Yu, L., Westland, S., Chen, Y., and Li, Z. 2021. Colour associations and consumer product-colour purchase decisions. *Color Research and Application* 46: 1119–1127, https://doi.org/10.1002/col.22659

Ziat, M., Balcer, C. A., Shirtz, A., and Rolison, T. 2016. A century later, the hue-heat hypothesis: Does color truly affect temperature perception?. In: Bello, F., Kajimoto, H., and Visell, Y. (eds), *Haptics: Perception, Devices, Control, and Applications*. Lecture Notes in Computer Science 9774. Cham: Springer, pp. 273–280, https://doi.org/10.1007/978-3-319-42321-0_25

4. Color and Healthcare

4.1 Introduction

Assuming that solar rays might have a therapeutic effect, Dr. G. L. Ponza, working in the psychiatric hospital in Alessandria in Italy, placed patients in rooms with colored-glass windows and colored walls. In around 1875, he reported that a patient, having been kept for a few days in a violet room, requested to be sent home since he felt cured. Ponza then experimented with other colors and found that red aroused melancholic patients, and blue calmed manic patients. Ponza's "color cure" stimulated experiments in asylums around the world, spanning colors such as yellow, black and brown. While the results were mixed, in the early twentieth century, many asylums implemented Ponza's "color cure" (De Young 2015).

This chapter will explore the role and use of colors in various medical and healthcare settings. First, it is noticeable that colors are associated with the names of diseases and clinical conditions, either because specific colorations are typical symptoms and signs of underlying pathology or because they evoke particular associations related to them. What is more remarkable is the diversified and widespread uses of color in the context of diagnosis. For a very long time, medicine has been more dependent on careful observations than on effective interventions. The first thing a physician should do, according to the founders of Western medicine, Hippocrates and Galen, is undertake a systematic and thorough inspection and examination of the patient. The body and its secretions can give important indications about possible diseases. The basic components of the human body, in concordance with the principal elements of the physical world, are colored so that changes in the bodily constitution are recognizable, as is the case with uroscopy. When medicine developed as a scientific enterprise in the sixteenth and seventeenth centuries, the epistemic status of color changed: instead of observing colorations of the body and its excretions and interpreting them within the framework of the humoral theory, colors are now used as tools to make anatomical and pathological structures visible within the body which can explain specific symptoms and disfunctions. Pigments and dyes are used as stains for cells, tissues and organs, and thus have resulted in numerous discoveries in histology, pathology, cell biology, and subsequently, bacteriology and genetics. Colors are actively employed to examine physiological processes with tests to determine the chemical properties of substances and fluids, such as the litmus test for acidity and alkalinity, and the Fehling's test for glucose in the urine. They are also helpful for distinguishing different types of microbes and cancer cells.

The third paragraph of this chapter will focus on medication. The tablets and capsules that are in use have a wide variety of colors and color combinations. It is generally assumed that the efficacy of medication is not only dependent on its chemical composition but on other factors, such as color. Diverse studies have investigated the medicinal effects of colored medication. However, there is a long history of using color itself as drug, which is the subject of the next section. In most cultures, pigments and medicines are prepared from the same source, namely natural substances, especially plants, animals, minerals and earth. They are often manufactured and sold by specific persons, i.e. grocers, druggists, chemists and apothecaries. Already in ancient civilizations, many pigments and dyes were used as medicines, despite the fact that several of them, particularly yellow and green pigments, are highly toxic. The traditional link between natural pigments and medication was revolutionized through the emergence of the science of chemistry in the nineteenth century, as discussed below. The synthesis of artificial colors was initially motivated by the search for new and effective drugs. Nonetheless, it first led to the production of synthetic dyes, such as aniline and alizarin, which introduced an array of new and bright colors. Upon further examination, the artificial dyes proved to have significant medical applications, not only as stains for diagnostic purposes but also as groundbreaking drugs. Many chemical companies that originally had concentrated on the dye manufacturing transitioned into pharmaceutical enterprises.

The subsequent section discusses color therapy, which emphasizes not the chemical properties of color but the belief is that color—or its perception—has curative power. An old example is the red treatment for smallpox. Advocates of chromotherapy argue that since colors have energy and vibration, they exert a healing effect on the human body. A contemporary variant is phototherapy, where the therapeutic benefit comes from not from color itself but from light. Exposure to blue or green light can have positive effects on pathological conditions.

The final section discusses the application of colors in the design of medical and care environments. There are multiple recommendations about optimal color choices for the interior design of hospitals and other healthcare facilities, emphasizing that colors and their associated meanings can evoke emotional and behavioral effects that aid patient recovery and enhance the well-being of staff and visitors of these facilities. However, research in this area is limited and conclusive evidence remains lacking.

4.2 Disease

The relation of colors to health and disease is first of all evident in the naming of diseases and pathological conditions. As pointed out in Chapter 1, some diseases and conditions are identifiable because they are associated with typical colors. A familiar example is jaundice, a yellow pigmentation of the skin caused by high levels of bilirubin in the blood due to metabolic, liver or biliary tract problems. The name is derived from the old French word *jaunice*, meaning yellowness. Another example of a disease named after a typical color is porphyria, a group of rare genetic disorders in which porphyrins accumulate as natural chemicals in the body, causing problems in the skin and nervous system. It was discovered because the urine of patients has a reddish, purple color (explaining the name of the disease, which is derived from the Greek term *porphyra*, purple). In these two examples, the observable color points to an underlying pathological condition within the body, and discoloration may be so characteristic that the disease itself is named after the color. Another example is rubella (or German measles), a viral infection with a light red rash starting on the face, spreading to the other parts of the body, and disappearing after three days. The name is derived from the Latin *rubellus*, diminutive of *ruber*, red.

Pink disease (acrodynia) was described in the beginning of the nineteenth century as a serious affliction of children, characterized by pink discoloration of the hands and feet. It was later discovered to result from mercury poisoning, caused by the ingestion of calomel (mercury chloride) in teething powders given to infants (Holzel and James 1952; Dathan 1954). In the same century, tuberculosis was referred to as the "White Plague", though the origin of the name is unclear. It is generally assumed to refer to the extreme pallor of infected people, but this is only the case in the terminal phase of the disease. In earlier phases, patients have a reddish appearance due to fever, and severe anemia was not a universal symptom (Weisse 1995). The attribution of the color white to tuberculosis emerged during an era when it was regarded as a "romantic disease," particularly affecting writers, composers and poets—artists with increased sensitivity and creativity. During this time, a consumptive appearance characterized by pale skin and red cheeks also became an aesthetic ideal, stimulating the use of arsenic cosmetics to make the complexion even paler. In this context, the color white reflected beauty and fragility but also purity, serenity and spirituality (Dyer 2017). Against this backdrop, calling tuberculosis a white disease does not so much refer to medical observations or pathological findings but rather evokes specific associations attributed to the specific color. As in the example of the "green disease" discussed in Chapter 1, the use of color in disease naming often serves a primarily symbolic function.

4.3 Diagnosis

That color is a sign of a healthy constitution is experienced in everyday interactions. We notice that someone is not feeling well or that something is wrong with someone because their skin color is different from usual. When someone is in good health, the Dutch use the expression "blaken van gezondheid", literary meaning "glowing of health," referring to a face with rosy cheeks. When the face is greenish, grey or pale, this indicates that a person is affected by illness or emotional distress. People generally associate face color with healthiness, an association confirmed in research that shows that skin redness (which is due to increased blood and oxygen in the skin) enhances the healthy appearance of faces (Stephen et al. 2009). It is therefore not surprising that color traditionally is used as a diagnostic indicator in medicine. Since therapeutic possibilities were very limited until the emergence of modern scientific medicine, physicians were primarily concerned with the correct diagnosis of an ailment and with prognostication of the course of the disease. That requires systematic inspection and observation of the patient. Hippocrates, the founder of ancient Greek medicine, argued that the diagnosis and treatment of diseases should begin with careful observation, assuming that every disease has a natural cause. He is famous for his description of the *facies Hippocratica*, a symptom of impending death. When the face is extremely pale and darkish, and there is no immediate improvement, the patient will die soon. Hippocrates is presumably also the first to describe "icterus" or jaundice as a separate condition, related to the pathology of the liver. He identified five types of jaundice which all have a yellow or greenish pigmentation of the skin. The diagnosis was based not only on the color of the skin but also on the color of urine and feces, and other determinants such as the season of the year. The first type of jaundice is characterized by a green skin and reddish sediment in the urine. The patient usually dies within fourteen days. The second type mainly occurs in the summer, when the heat of the sun was thought to move the bile, causing the yellowish color of skin, urine and stools. The prognosis was bad. Nowadays, the first type correlates with hepatic jaundice due to hepatitis A and acute cholangitis, while the second type is post-hepatic jaundice as a result of pancreatic cancer (Papavramidou et al. 2007).

The significance of careful and systematic observation in medicine continues to be emphasized by Hippocrates' successors. In the first century, Aulus Cornelius Celsus, in his extensive *De Medicina*, advocates observation of disease development in order to identify prognostically bad symptoms. For example, when a patient has acute fever, vomiting of phlegm or bile is dangerous, and worse still if it is green or black (Celsus 2021, Book II. 4). When the color of a patient is black or very pale, future death can be expected. When there is an infection of the lungs and the patient starts spitting, the more mixed and less distinct the colors of the sputum are, the worse; but the worst of all the colors is black (Celsus 2021, Book II. 8). Due to his detailed description of clinical signs and symptoms, especially of fevers, Celsus is regarded as the first to have identified the cardinal signs of inflammation: *rubor* (redness), *calor* (heat), *tumor* (swelling) and *dolor* (pain). In the second century, Galen of Pergamon brings the Hippocratic tradition of systematic observation to another level not only with accurate descriptions of clinical cases but also with experiments on animals to understand the anatomy, physiology and pathology of the organism (Cosans 1998). As a prolific writer, Galen had a major influence on medical thinking and practice, continuing into the seventeenth century. One of the reasons is that he provides a comprehensive theoretical framework for diagnosis, therapy and prevention. Elaborating the theory, already proposed by ancient Greek philosophers and Hippocrates, that the human body is composed of four fluids, Galen argued that this humoral constitution is analogous to the fundamental elements of the physical world (black bile/earth, phlegm/water, yellow bile/fire and blood/air) as well as to basic qualities of hot, cold, wet and dry. He furthermore associates the humors to different temperaments or personalities, respectively melancholic, phlegmatic, choleric and sanguine. By invoking equivalence between elements internal and external to the body, this theoretical framework relates the human to the natural, the micro-cosmos to the macro-cosmos, but also the physical body to psychology. It explains the occurrence of health (when the humors are balanced) and disease (when there is an excess or lack of humors), and provides indications for appropriate medical intervention (Singer 2021). This framework is deeply influenced by colors, since the humors are recognizable as black, white, yellow or red.

The idea that the color of the body or body parts gives important clues about health and disease has especially been applied for diagnostic and prognostic purposes. An example is the practice of uroscopy (Neuburger 1937).

The Hippocratic writings describe how the color of urine and urine sediments can provide indications about pathology and the course of a disease. On the basis of the different qualities of urine, Celsus differentiates diseases of the bladder, kidneys and other organs. In the Middle Ages, uroscopy became so important that it was considered the actual basis of medical practice. Because health and disease were believed to be determined by the bodily humors, the color of urine was thought to indicate the dominance of a particular humor, allowing for the condition of the whole body to be diagnosed. For this purpose, colored illustrations (so-called urinary discs) were widely used. The urine glass became the chief symbol of medicine, and physicians are often presented in works of art holding this glass. From the seventeenth century onwards, visual inspection of urine colors was gradually replaced by chemical and physical examination.

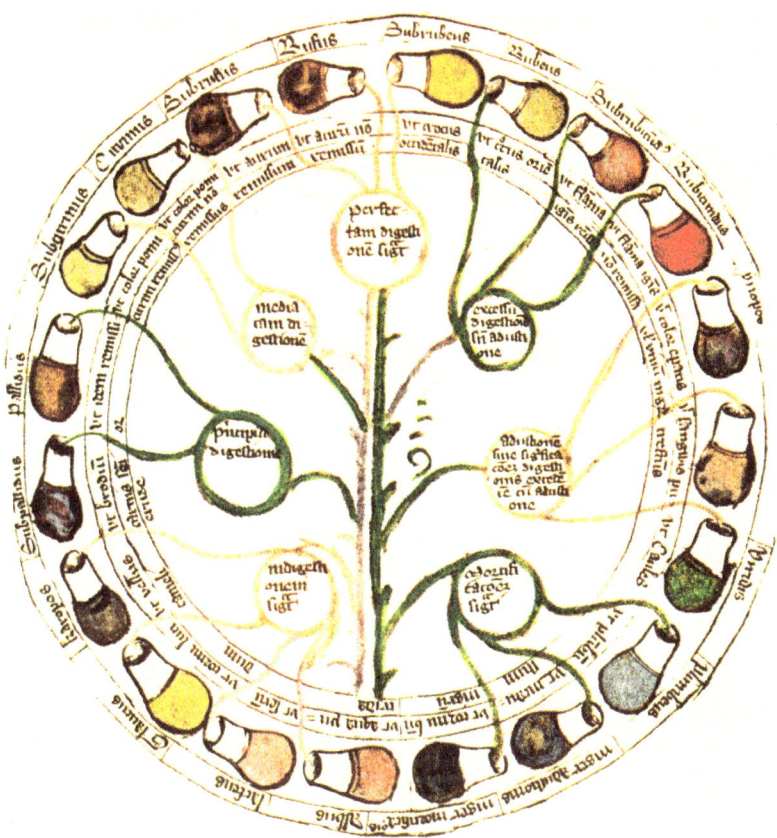

Fig. 4.1 Johannes de Ketham, *Fasciculus Medicinae* (1491). Uroscopy chart relating the color of urine to bodily constitutions and ailments. Wikimedia, https://commons.wikimedia.org/wiki/File:Fasciculus_Medicinae_1491.jpg#/media/File:Fasciculus_Medicinae_1491.jpg, public domain.

In contemporary medicine, color is still regarded as an important sign of health and disease. Discoloration of the skin, for instance, can be a normal phenomenon of ageing but may also indicate pigmentation disorders such as albinism (absence of melanin), vitiligo (pale white patches) or melasma (dark skin due to hyperpigmentation). In some cases, it reveals an underlying condition that needs medical attention, such as Addison's disease (with darkening of the skin), dermatitis (redness of the skin), psoriasis (red patches) and skin cancer (Bottaro 2023). Another example is the color of the tongue. Normally it is light to dark pink. A black hairy tongue means that the papillae are too long and the tongue has become brown or black due to keratin. It may be the result of poor oral hygiene, smoking or certain medications. White patches on the tongue indicate yeast in the mouth (thrush), usually because of underlying conditions such as diabetes or HIV, but also potentially leukoplakia, a precancerous condition. A bright red tongue (called a "strawberry tongue") may indicate vitamin B12 deficiency or scarlet fever. A blue tongue means cyanosis, resulting from a lack of oxygen in the blood, and may signal blood disorders, blood vessel disease or kidney disease (Cleveland Clinic 2023).

Although it remains important to observe the patient's body and its appearance, the emergence of scientific medicine has significantly changed the relevance and use of colors in diagnostics. Following the development of anatomy in the sixteenth century (with the work of Andreas Vesalius) and physiology in the seventeenth century (the discovery of the circulation of blood by William Harvey), humorism as an explanatory theory was gradually replaced with a mechanistic interpretation of medical science regarding the human body as a complex biological machine which can be examined and described with the methods of natural sciences (Broadbent 2019). The new discipline of pathology identifies and classifies the lesions and bodily changes characteristic of diseases while the goal of clinical medicine is to relate signs and symptoms to underlying conditions and changes in the interior of the body. Careful observation of the bodily surface will therefore provide clues to pathological processes within organs, tissues and cells, explaining the symptoms and signs of the patient. In this scientific medical perspective, color obtains a different role. Instead of visually examining and interpreting the colors of the body and its excrements, they are regarded as indicators of underlying mechanisms and can be used to highlight changes in these mechanisms. Rather than being simple surface phenomena that either hide and disguise or reveal and accentuate the objects they cover, they can now be used as diagnostic tools to expose the fundamental biological, chemical and physical structures that are responsible for symptoms and disfunctions, and can therefore be employed as indicators and markers to make the normal and abnormal functioning of the bodily organism visible.

One of the first uses of color for this purpose was the litmus test. Traditionally, pigments produced by lichens were used as dyes for clothing but a mixture of these dyes, absorbed on a filter paper, can be used as a pH indicator. Already around 1300, pigments were used to study acids and bases. Called "litmus" from the Old Norse word for "moss used for dyeing," a blue paper becomes red in an acidic medium and a red paper blue in a basic medium.

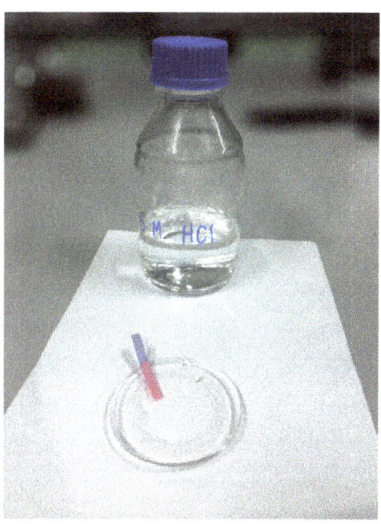

Fig. 4.2 Red litmus paper reacts with hydrochloric acid in litmus test. Photo by Kanesskong (2016), Wikimedia, https://commons.wikimedia.org/wiki/File:The_result_of_red_litmus_paper.jpg#/media/File:The_result_of_red_litmus_paper.jpg, CC BY-SA 4.0.

The famous chemist Robert Boyle in 1664 argued that if dyers could change the colors of a lichen extract by adding acid or alkali, the extract could be used to indicate the acidity or alkalinity of unknown substances (Brock 1993, 178). Until the nineteenth century, litmus was almost exclusively produced in the Netherlands (as "lakmoes"). Since the test was simple and decisive, it came to be used as metaphor, referring to a single factor that may reveal the character and opinions of somebody. The Secretary-General of the United Nations, for example, in September 2023 called the response of the international community to the devastating floodings in Pakistan a "litmus test" for climate justice (United Nations 2023).

Apart from showing functional qualities of substances (such as acidity), colors (as dyes and pigments) were increasingly used to draw attention to the fundamental makeup of biological entities. Antonie van Leeuwenhoek (1632–1723), who was the first to use self-fabricated microscopes for the study of minute organisms and cells, colored his specimens with existing dyes such as saffron, madder and indigo (Javaeed et al. 2021). Pigments which have been long available, such as carmine, a bright red colorant, were increasingly used to make miniscule anatomical structures visible. When new pigments were discovered, for example, Prussian blue in the early eighteenth century, they were also used for the same purpose. As soon as the first synthetic dyes were produced in the nineteenth century (e.g. mauveine and aniline), they were rapidly introduced in microscopic studies. Paul Ehrlich (1854–1915) applied them to identify different types of white and red blood cells and to stain various species of pathogens. Colors therefore not merely made biological structures and components visible, but experimenting with various dyes and staining techniques brought hitherto unsuspected anatomical and pathological objects and systems into the light. New discoveries were made with staining techniques. In 1873, Camillo Golgi published the first picture of a nerve cell, using a silver staining technique. Dyes are furthermore used in practical diagnostics. For example, Fehling's solution introduced in 1849 enabled physicians to discover sugar in the urine, and thus to diagnose diabetes. When its bright blue color turns red-brown or orange the urine contains glucose.

Today, microscopic study of cells and tissues would be unthinkable without a wide variety of staining techniques (Alturkistani et al. 2016). As the examples illustrate, the use of pigments and dyes, and thus color, has contributed much to the progress of scientific disciplines such as histology, pathology and microbiology, as well as to the advances in medical practice due to enhanced diagnostic capabilities. Dying and staining are primarily employed to make cells, tissues and their components visible under the (light or electron) microscope. They can then be examined and the underlying mechanisms of diseases explored. In present day healthcare, for example, dyes in ophthalmology provide a better view of the ocular surface and the vasculature of the retina. Colors make cancer visible; they are indispensable in histopathology, imaging and molecular diagnostics; in cardiac catheterization contrast dyes are injected to exhibit blockage in the cardiac arteries; and dentists have diagnostic tests using dyes to detect and remove caries. Secondly, dying and staining techniques are applied to differentiate between microorganisms, cells and tissues, and to identify what is normal and abnormal. This is not only important for an adequate diagnosis but also for proper medical intervention. The staining method invented in 1875 by Hans Christian Gram, for example, allowed for a distinction to be made between different types of bacterial infection.

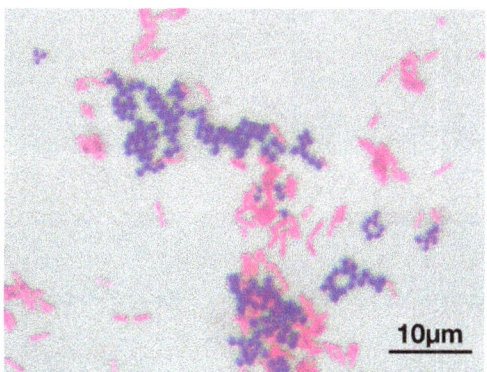

Fig. 4.3 Microscopic image of a Gram stain of mixed Gram-positive Staphylococcus aureus (purple) and Gram-negative Escherichia coli (red). Image by Y tambe (2010), Wikimedia, https://commons.wikimedia.org/wiki/File:Gram_stain_01.jpg#/media/File:Gram_stain_01.jpg, CC BY-SA 3.0.

Learning whether bacteria are gram-positive or gram-negative determines which treatment is appropriate for an infection. A well-known illustration of this second point is the Pap smear, used in cervical screening. George Papanicolaou spent years testing more than four hundred combinations of imported and domestic stains in order to distinguish precancerous and cancer cells from benign cervical cells. In his laboratory, almost every day a new stain or staining procedures was tested to the despair of his assistants; eventually, in 1942, he identified a combination of five dyes to make visible cervical cancer (Chantziantoniou et al. 2017). The smear is now widely used in screening programs around the world and may have significantly reduced cervical cancer deaths.

4.4 The Color of Medication

Color historian Michel Pastoureau recounts how he grew up in his mother's pharmacy in Montmartre, Paris in the 1950s. Colors are crucial to ordering and classifying in this setting and to identifying various materials. White dominates the boxes with medication; it refers to hygiene, science and benefits. Blue is used for sedatives, sleeping pills and anxiolytics; yellow or orange for the opposites: tonics, vitamins and everything that provides strength. Beige and brown are reserved for drugs for digestion or the digestive system, while black is avoided. Green is sparsely used, usually for herbal and alternative medicines, but it is also the color of the sign at the exterior of the pharmacy. Since the Middle Ages, a green cross is the emblem of apothecaries because most medicines were derived from the plant world (Pastoureau 2010).

When we look at the medication we are using, it will be noticeable that pills and tablets are often white, although many drugs, especially capsules, have a range of colors. Numerous studies have focused on the effects of colors of medication. The underlying idea is that how patients respond to drugs is not only determined by their chemical composition but also by other factors. What they look like (preparation form, shape, size and color) may generate certain expectations concerning action and strength. Capsules, for example, are perceived as more powerful than tablets (Buckalew and Coffield 1982). Studies with placebo drugs (that have no active component) indicate that psychological effects are produced by colors: subjects classify red and yellow placebos as stimulants, and blue placebos as depressants (Jacobs and Nordan 1979). Specific colors, and the meanings ascribed to them, apparently determine perceived potency of drugs and expectancies for drug action. Red and black capsules are found to be associated with a presumption of high potency whereas white is perceived as weak (Sallis and Buckalew 1984). Another study identifies the pharmacological effects of four colors. White is associated with analgesic action (most pain-killing medication is white), lavender with hallucinogenic effect, orange and yellow with stimulant-antidepressant action. For dark red, dark blue and light green, no specific associations can be found (Buckalew and Coffield 1982). A review of the literature concerning the color of oral drugs that affect the central nervous system confirmed that red, yellow and orange are associated with a stimulant effect, blue and green with a tranquillizing effect (De Craen et al. 1996). A study testing the effect of oxazepam tablets in three different colors (red, yellow and green) concludes that symptoms of anxiety improved most with green, and depressive symptoms improved most with yellow tablets (Schapira et al. 1970). The impact of the color of pills on the clinical response is also rather consistent across continents. Amawi and Murdoch (2022), for example, find that in various countries the color of pills exhibit similar effects on the perceived efficacy in various categories: for sedatives, blue and white; for stimulants, red; for anti-anxiety drugs, blue and white; for pain relief, white; for antacid effects, yellow and white; and for hallucinogenic effects, yellow and red. Only green does not have strong associations to any of the efficacies studied. These findings correspond very well with studies on the different meanings attributed to colors, discussed in the previous chapter: blue has connotations of calmness, red of activity and excitement, white of purity and tranquility, yellow of energy and warmth.

It is reasonable to assume that medication has always been colored. In the tenth century, Avicenna gave instructions for preparing medicines and advocated the practice of gilding and silvering pills. In nineteenth-century England, white pills (with lactose) were distinguished from colored ones (with powdered liquorice root) and in the Parliamentary debate on the National Insurance Bill in 1911, reference is made to the fact that medicines are colored (Anderson 2005, 210, 85). Pills were also advertised for their colors, for example Pink Pills for Pale People, used for a wide range of diseases (Anderson 2005, 233). The relation between medication and color is not surprising because, traditionally, many pigments were also used for medical purposes and both were often manufactured by the same people. For centuries, medicines, like pigments and dyes, were prepared from naturally occurring materials, mostly plants but also minerals and animal substances. Hippocrates, Celsus and Galen emphasized the importance of plants for medical treatment. Medicinal products came in many forms: powders, ointments, syrups, lotions, elixirs, tinctures, enemas and pills. The majority were liquid preparations. Still at the beginning of the twentieth century, in hospitals in England around sixty percent of medicines took liquid form, manufactured on the premises (Anderson 2005, 137). Pills were probably made to deliver measured amounts of a medicinal substance and to overcome the unpleasant taste of liquids. The first references to pills are found in ancient Egyptian papyruses (Mestel 2002).

Historically, pills were handmade. After powdering, the solid active ingredients were mixed with bread dough, honey or oil to make a stiff mass, and then formed into roughly spherical substances using the fingers. These pills were normally larger than the later pills of the nineteenth and twentieth centuries. They were usually sugar-coated. In the 1840s, machines capable of producing pills became available, as well as machines that could compress medicinal substances into tablets. In the 1870s, gelatine capsules were invented. The emergence of the synthetic dye industry in the same century produced an avalanche of new dyes which could be used for coloring medicines.

Nowadays, capsules and tablets can have thousands of color combinations. Coloring medication is justified with several arguments. It helps consumers to recognize medication better and to distinguish them from each other so that the risk of accidental poisoning is diminished. For pharmaceutical companies, coloring most often serves marketing purposes; it helps to identify brands, especially when pills are available over the counter. For example, Nexium is advertised as the purple pill. A third argument refers to the supposed power of colors. If colors have particular meanings and raise specific expectations about efficacy, then responsivity of patients to medication will be better when its color corresponds with its perceived intended effects (i.e. red for speedy relief, and blue for sleep and calmness). Coloring medication may therefore have implications for the compliance and response of patients. A drug will be more effective if the meaning attributed to its color corresponds to the expected efficacy.

4.5 Pigments as Pharmaceuticals

Until the nineteenth century, pigments were made from plants, animals (particularly insects and shellfish), minerals (e.g. copper) and earth (e.g. ochers). The surrounding world provides a multitude of colored materials, and many artists fabricated the pigments they used for painting themselves. Dutch painter Johannes Vermeer (1632–1675) ground many of his pigments at home, ordering the raw materials at the nearby apothecary shop (Jelley 2017). Grocers, druggists and chemists manufactured and sold pigments, together with spices, medicine and often food. The first reference to the pharmaceutical profession appears in the sixth century, when it was stated that doctors' prescriptions were carried out by a so-called *pigmentarius*, who sold pigments, spices, drugs and dyestuffs. However, a distinct separation between medicine and pharmacy in Western Europe did not develop before the thirteenth century (Boussel, Bonnemain and Bové 1983, 88). In the sixteenth century, a special profession of color manufacturer and dealer emerged (Gage 2013). This became even more important with the invention of synthetic colors; manufacturing paint was no longer a matter of grinding solid pigments but the outcome of chemical synthesis. Artists became dependent on special color shops, such as Julien Tanguy's store, with Vincent van Gogh and Paul Cézanne as regular clients (Ball 2009).

Since Antiquity, numerous pigments have been used as medication. Since green is regarded as a soothing and relaxing color, eye ointment made from pulverized emeralds was used in Roman times. A popular drug was mummy brown, a pigment made from the grounded flesh of mummies, administered as ointment or potion. Plinius recommends it as toothpaste, Francis Bacon as drug to stop bleeding, and Robert Boyle as a remedy for bruising (St Clair 2016). The plant *crocus sativus* is traditionally the source of the yellow-orange pigment saffron, but it also has medical uses as an aphrodisiac, painkiller and stimulant (Eckstut and Eckstut 2013). Another yellow pigment, gamboge, was derived from trees in South-East Asia and collected in bamboo canes. First brought to Amsterdam, it was studied by the sixteenth-century botanist and physician Carolus Clusius and sold as a pharmaceutical against rheumatism and scurvy, and in small doses as a laxative. A legendary example is kohl, the black cosmetic employed in Ancient Egypt as eye liner and to prevent eye infection, still used today. Analysis of old samples of kohl shows that it contains natural pigments (such as black sulfide of lead, sometimes mixed with grounded pearls, coral or emerald for a shining effect) but also chemicals, particularly chlorides of artificial lead (Riesmeier et al. 2022). These substances stimulate the skin around the eyes (producing more nitrogen oxides) which may reduce the risk of eye infections (St Clair 2016). The analyses demonstrate the diversity of materials and recipes to produce kohls but, even more, the efforts of manufacturers of pigments to create medicinal preparations with the use of chemicals, thus producing the first synthetic pigments (Delamare and Guineau 2000).

The disadvantage of using pigments as drugs is that many are known for their toxic effects. Yellow pigments, for example, are very poisonous when used as medicines (Pastoureau 2019). This is dependent on the dosage. Gamboge, for example, is fatal if taken in large quantities. Toxicity can also result from the pigment's composition as a medicine. Orpiment, a naturally occurring mineral widely used in Ancient Egypt, contains sixty percent arsenic and is deadly if ingested. Recommended as a remedy against hair loss and to repel insects, its toxicity is noticed, for example, by the Greek geographer Strabo in the first century, who reported that only criminals were employed to mine the mineral (Delamare and Guineau 2000). Another ancient pigment commonly utilized in various cultures is lead white. It is an essential paint for artists but is also applied in houses as wallpaper and as a cosmetic. It is not expensive and relatively easy to fabricate, but the production process is dangerous due to toxic sulfurous fumes. When used as powder to whiten and smooth skin, lead white frequently caused lead poisoning, especially in babies exposed through breast feeding (St Clair 2016). In the nineteenth century, it was replaced by synthetic whites (e.g. titanium dioxide) which significantly reduced cases of lead intoxication.

While several natural pigments were known to be poisonous, later, in the search for new pigments, toxicity arose due to the mixture of natural substances with chemicals. For example, Naples yellow, invented in the seventeenth century and prepared artificially with a mixture of lead and antimony, is extremely toxic. It has been suggested that Vincent van Gogh's mental illness was related to his preference and frequent use of this yellow paint (Delamare and Guineau 2000, 137). A notorious example of a dangerous pigment is Scheele's green. The color green is not laborious to obtain from plants, roots, leaves and flowers, but in paintings it is difficult to stabilize and as a dye it does not attach well to fibers. In the lengthy search for better green pigments, Carl Scheele discovered in 1775 copper arsenite, a pea-green pigment that became used in paintings but also in textiles and wallpapers, as well as in confectionary and pastries. It did not take long before it was observed to cause symptoms such as nausea, vomiting, diarrhea, rash and lethargy, particularly when arsenic particles become airborne in dry settings. While Scheele knew that the pigment was poisonous, its use was never prohibited; it simply fell out of favor and became obsolete with the introduction of new synthetic greens (St Clair 2016). Several green paints became associated with poison. Since it was the preferred color of Napoleon who had all his rooms in St. Helena painted in green, for a long time, theories circulated that his death was due to arsenic intoxication. In the 1950s, the American ambassador in Rome became seriously ill when she stayed in her residency. At first it was assumed that she was the victim of Soviet agents, but medical analyses showed that she had arsenic poisoning. It appears that the arsenic-based paints in the bedroom and dining room of the fifteenth-century residential villa was flaking and that the ambassador had been breathing arsenated fumes and intoxicated coffee for months (Time 1956). In 2022, our attention was drawn to toxic books. In the mid-nineteenth century, publishers produced books bound in cloth colored with emerald green, a combination of copper acetate with arsenic trioxide and invented in 1814. People who handle these books nowadays frequently may inhale or ingest arsenic particles and should take special precautions when they review them (Brower 2022).

The toxic effects of natural as well as synthetic pigments and dyes confirmed the traditional suspicion that at least some colors, especially bright ones, could be dangerous. Particularly green is associated with poison, not only in the case of pigments, but also when new substances are identified. When Marie and Pierre Curie discovered radium in 1898, it was reported to emit a greenish glow. Since then, radioactive material is associated with the color green (which is a myth, since radioactivity is not visible to the human eye). However, the myth is not without any grounding. If radium is added to paints containing phosphorescent copper and zinc sulfide, they acquire a greenish luminosity, particularly in the dark. Companies started to produce items such as luminous clocks and watches with this "radioactive green." This paint was used until the 1970s, despite knowledge of the dangers of radiation (Lu 2020).

4.6 The Pharmaceutical Revolution

Fig. 4.4 Jerry Allison, *William Henry Perkin—Pioneer in Synthetic Organic Dyes* (1980). Science History Institute. Perkin (center) in his laboratory examines test dying of silk taffeta with mauve aniline dye. Wikimedia, https://commons.wikimedia.org/wiki/File:William_Henry_Perkin-_Pioneer_in_Synthetic_Organic_Dyes_-_DPLA_-_3acf6c4043b0ea3ee1044c835092c5ec.jpg#/media/File:William_Henry_Perkin-_Pioneer_in_Synthetic_Organic_Dyes_-_DPLA_-_3acf6c4043b0ea3ee1044c835092c5ec.jpg, CC BY 4.0.

The use of the same substances as pigments as well as medication completely transforms in the nineteenth century, due to the emergence of the science of chemistry. The groundbreaking event was the discovery of the new color mauveine by William Perkin (1838–1907) in 1856 (Garfield 2000). As a young disciple of the chemist August Wilhelm Hofmann (1818–1892), Perkin experimented with coal tar in search of a synthetic alternative for quinine, the first known medication against malaria. Quinine is extracted from the bark of the South American cinchona tree, but is expensive and in short supply. Coal tar is the byproduct of coal gas, and abundantly produced by gas lighting, which became a popular mode of public lighting in many cities during the nineteenth century. Perkin discovered accidentally that this waste product could be used for the synthesis of dyes, in his case a beautiful purple that colored silk and wool. He discovered how to make color from coal, and started a commercial enterprise to produce artificial colors, demonstrating how chemical research could have significant practical and economic implications.

The basic substance in Perkin's dyestuff is aniline, first extracted from the natural dye indigo but later from coal tar (Ball 2009). At the beginning of the nineteenth century, scientists succeeded in isolating organic compounds from plants and animals in order to explain the pharmacological effects of natural medicines. The first of these so-called alkaloids is morphine, isolated from opium in 1804, followed by atropine from nightshade plants (1819), quinine from the cinchona bark (1820), caffeine from coffee beans (1820) and nicotine from tobacco (1828), among many others. It has been known for some time that aniline, extracted from coal tar, is associated with color-producing reactions but scientists such as Hofmann were primarily interested in it as a possible precursor for the chemical synthesis of quinine, and perhaps other drugs. Due to Perkin's discovery, experiments with aniline were focused on synthetic dye production, introducing new colors such as fuchsine or magenta (deep red) due to commercial interests and the growing market for new and inexpensive dyes in the textile and fashion industries (Garfield 2000). In the early 1860s, aniline dyes were also used to stain cells and tissues in the human body. They make, for example, structures visible in the cell nucleus: so-called chromosomes ("colored bodies"). The increasing demand for aniline dyes boosted the development of major chemical companies, especially in Germany—where Friedrich Bayer (1825–1880) in 1862 began the production of aniline reds and blues, and where in 1865 the Badische Anilin and Soda Fabrik (BASF, later IG Farben) was established—and in Switzerland, with the emergence of Ciba, Geigy and Sandoz as major manufacturing enterprises (Boussel, Bonnemain and Bové 1983). As a result of artificial synthesis, the traditional trade of natural dyestuffs collapsed. The indigo industry in Britain which imported blue dye from India imploded after 1880, when Adolf von Baeyer (1835–1917) produced artificial indigo (for which he received the Nobel Prize in Chemistry in 1905), shifting the global trade balance and providing Germany a worldwide monopoly on this dye. Another example is the madder plant (*Rubia tinctorum*) an ancient source of crimson red dyes. It was intensively produced in regions such as Provence and Holland. In 1827, the coloring substance in the plant was identified (alizarin) and synthesized in the laboratory in 1868. Synthetic alizarin became the color of the 1860s, brighter and less expensive than the natural pigment, causing the near extinction of traditional production and making Germany the market leader. In the beginning of the twentieth century, Germany produced eighty-eight percent of all colorants in the world (Delamare and Guineau 2000, 110).

The discovery and production of artificial colors firstly demonstrates the emergence of chemistry as a practical science. The use of natural pigments gradually disappeared, now that synthetic dyes could be produced through chemical synthesis. While chemistry had a low status before the eighteenth century, and production of colors was the business of dyers and apothecaries without much theoretical knowledge, chemical research in the following century not only highlighted important commercial applications but also the idea that colors could be rationally synthesized. This idea as such was not new. Color and color technology have closely been related to the traditional practice of alchemy, aimed at the transformation of primary qualities of substances, and thus the replacement and improvement of natural processes. An ancient example of a manufactured pigment is vermillion, a red-orange color made from mercury and sulfur (Gage 2013). For the physician Paracelsus (1493–1541), who studied the use of chemicals and minerals for medical purposes, alchemy was the art of transformation. Alchemists noticed that changes of materials are accompanied by alterations in color, usually in four stages, from melanosis (blackening), leukosis (whitening), xanthosis (yellowing) to iosis (reddening), reflecting the movement of matter through air, earth, fire and water (Riley 1995). Experimenting with different kinds of matter, alchemists were driven by practical motivations, focused on assisting and improving the work of dyers, metallurgists, pharmacists and physicians. Because it stimulated empirical research, alchemy positively contributed to the development of chemistry (Brock 1993). It furthermore expressed the belief that scientific knowledge can be used to improve the human condition, because the natural world is not immutable but can be perfected if its constitution is known. Chemistry, as it developed from the eighteenth century as the science of analysis and synthesis of materials, brought this belief to fruition, analyzing substances into their elements and synthesizing substances from their elements. Chemical synthesis ultimately replaced natural dyes and pigments, delivering new, brighter and cheaper products.

Fig. 4.5 Paul Ehrlich, c. 1910. Photographer unknown. Wikimedia, https://commons.wikimedia.org/wiki/File:Paul_Ehrlich,_c._1910.jpg#/media/File:Paul_Ehrlich,_c._1910.jpg, public domain.

Secondly, the chemical production of artificial colors shows the close connection between the chemical industry and discovery of new medication. Chemical research was motivated to find medical applications and found dyes, as the example of Perkin illustrates. The initial purpose of tar research was the fabrication of drugs, not colors. But while working with dyes, medical effects were noticed and new applications discovered (Garfield 2000, 185). Synthesis of dyes not only led to new discoveries in cell biology and bacteriology but also to successful treatments of illnesses (Gage 2013). A first example is phenol (carbolic acid), isolated from coal tar in 1841. It can be converted into aniline, but more importantly, it has effective disinfectant properties. Around the same time, the sanitarian movement and public health initiatives addressed unhealthy and disease-producing conditions in the human environment, hence there was a need to improve hygiene. Phenol was then used as an antiseptic, for instance as carbolic soap and local anesthetics, and from 1865 was applied by Joseph Lister (1827–1812) in the surgical treatment of wounds, making surgery much safer (Brock 1993). While willow bark extracts were long known for their analgesic properties, in 1838 the compound salicylic acid was isolated from the bark, and in 1860 this substance could be synthesized from phenol. This led to the production by the Bayer company of one of the most successful drugs, in 1897: aspirin (Ball 2009). A second example of the relation between synthetic dyes and medication is the work of the earlier mentioned Paul Ehrlich.

Supported by the dye industry of Hoechst, Ehrlich was one of the first to use aniline and alizarin dyes in histology and bacteriology in the early 1870s. His studies with the aniline dye methylene blue inspired Robert Koch to use this dye to detect the bacillus that caused tuberculosis. Ehrlich's most important contribution was noticing that dyes not simply colored cells and tissues, but adhered to some substances and not to others, because they combined with them and produced a chemical reaction, and that some dyes killed the microorganisms that absorbed them (Garfield 2000). This led him to the concept of chemotherapy: synthetic compounds could be working as drugs that destroy specific microorganisms and thus could cure diseases. In 1909, he discovered that an arsenic-containing drug killed the bacteria that caused syphilis. This drug was marketed by the Hoechst company a year later as Salvarsan. Because a chemical compound could be synthesized that selectively targeted a microbe responsible for a disease, Ehrlich introduced the concept of the "magic bullet," implying that scientific research could develop effective pharmacological products to eliminate diseases. A third example of the medical benefits of dyes and pigments is the development of the first sulfanilamide drug against streptococcal and staphylococcal infections. The German biochemist Gerhard Domagk (1895–1964), working at IG Farben, systematically tested thousands of synthetic dyes in the 1920s and found that the red dye Prontosil operated as an antibacterial agent useful for the treatment of pneumonia and puerperal fever.

To conclude, the investigation of coal tar as the apparently worthless residue of gas lighting generated the aniline dye industry. The growing elucidation and understanding of the chemical constitution of substances during the nineteenth century enabled the production of numerous new colors. The enormous demand for colors in the textile and fashion industry made it attractive for chemical researchers to manufacture ever more synthetic dyes. As chemical knowledge and technology advanced, deliberate manipulation and production of natural materials became feasible, making traditional pigments and dyes obsolete; color is no longer found in nature but produced in the laboratory. Synthetic colors have become so influential and common that the artificial supplanted the natural. As a result, the chemical dye industry emerged as a paramount commercial sector of society, especially in Germany, France and Great Britain. This had major repercussions for other social activities, for example, providing inspiration to new movements in the art of painting such as impressionism. The most significant effect, however, was in the field of pharmacology and medicine, in which numerous new and effective drugs became available. This pharmaceutical revolution was born in the research and production of synthetic dyes. Many pharmaceutical companies had their origin in chemical dye industries. The search for colors ultimately produced the therapeutic arsenal of contemporary medicine.

4.7 Color Therapy

While color as chemical matter (as pigments, dyes and synthetic substances) has instigated the development of numerous drugs, another line of thinking emphasizes that colors as visual sensations, or at least the perception of colors, may have a beneficial effect, as demonstrated in the example of the Ponza color cure mentioned at the beginning of this chapter. Colors, while influencing the efficacy of medication, may also impact the well-being and emotional status of patients who do not use drugs. Studies indicate that colors can affect pain perception, with certain hues intensifying or alleviating discomfort. For instance, red has a stimulating effect and has been shown to make pain feel more intense, in contrast to green and blue. White, associated with purity, cleanliness and hygiene, is recognized for its pain-reducing and sedative properties (Wiercioch-Kuzianik and Babel 2019). From here, it is a small step to regard colors as therapeutic.

In the history of medicine, smallpox was a deadly viral disease. After a global campaign, it was finally eradicated in 1980. As a scourge that affected humanity for millennia and with a fatality rate of twenty-five percent, it probably killed more people than the plague. In numerous cultures around the world, it was accepted that the color red protected against this disease and could even heal it (Hopkins 2002). The red treatment was first proposed by Avicenna in the twelfth century. The underlying idea was that red had exciting and warming properties, such that wrapping a patient in red clothing would draw the bad humors of smallpox to the surface of the skin, where they can be excreted. Centuries earlier, red treatment was already practiced in Japan, China and India. Red materials were brought into the sickroom, patients were covered in red, and given red drinks. In these countries, deities to protect against smallpox were imagined as red-colored. In nineteenth-century Europe and the United States, physicians developed erythrotherapy, claiming that red light would reduce the formation of smallpox pustules. While clinical studies were set up to evaluate this method, conclusive benefits could not be found.

The belief in the healing powers of colors has stimulated various forms of color therapy. Advocates of chromotherapy have developed theories to specify the action of colors upon different organs and systems of the body. When color is primarily regarded as a physical phenomenon, it can be supposed that each color has a distinct wavelength and thus vibration, affecting the body and specifically its chemical constitution. Diseases can therefore be healed by specific vibrations, and for each disease a particular color can be used. Color therapy therefore is vibrational healing (Klotsche 1992). For example, since red has the longest wavelength and the lowest vibratory rate, it stimulates and activates digestion and the liver. Blue on the other hand has a catabolic effect and reinforces the immune system, since it has the shortest wavelength and the highest vibratory rate. While the medical effect of color is explained in physical and chemical terms, most theories also appeal to their psychological and emotional influence. They repeat and elaborate the generally known associations of colors and specify the applications in healthcare. Warm colors such as red and orange are exciting and should not be used with fevers and inflammations. It is also argued that they rejuvenate the human body by purifying the blood

(Klotsche 1992, 51). Orange boosts the energy in the lungs and stomach, while yellow stimulates the sensory and motor nervous system. Cool colors such as blue and violet relieve fevers and pain. Green is supposed to encourage the activity of the pituitary gland and thus to restore the balance between anabolism and catabolism in the body (Anderson 1990, 19). It can also be used for patients with mental illnesses. Light green beds, for example, help to calm disturbed patients (Anderson 1990, 32). Promotors of chromotherapy argue that the advantage of this approach is that it leaves no harmful residues in the body, in contrast to medication (Anderson 1990). While color as matter (i.e. as synthetic dyes developed into drugs) has generated chemotherapy and drugs that are ingested or injected in the body, color as visual sensation may also produce physical and chemical effects within the bodily constitution, without possible adverse effects.

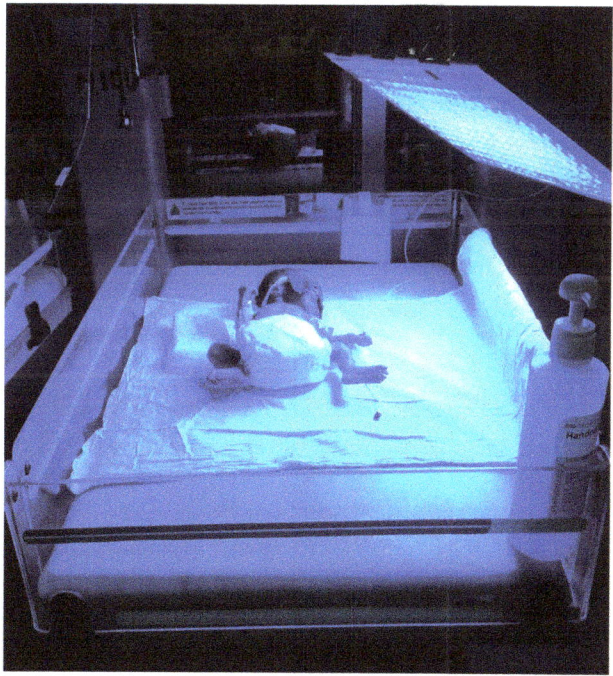

Fig. 4.6 Phototherapy of neonate for jaundice. Photo by Vtbijoy (2013), Wikimedia, https://commons.wikimedia.org/wiki/File:Phototherapy.jpg#/media/File:Phototherapy.jpg, CC BY-SA 3.0.

In turn, in many cultures throughout time, sunlight was believed to have healing powers. Physician and scientist Niels Finsen (1860–1904) is regarded as the father of modern phototherapy because he used radiation with light for the treatment of diseases (for example, red light to treat smallpox) with the argument that certain wavelengths of light have beneficial effects; he received the Nobel Prize in Physiology or Medicine in 1903 for his work. Since then, light therapy (phototherapy or heliotherapy) is applied for a variety of disorders. Natural or artificial light is used to improve skin conditions such as eczema and psoriasis (with ultraviolet light) and for mood and sleep disorders. Phototherapy is commonly applied when babies have an excess of bilirubin. Newborns with severe jaundice are exposed for a short time to blue light. In 1956, it was discovered that infants lose their yellowish color in sunlight. Short wavelengths of light (perceived as blue) alter the molecular structure of bilirubin and make it water soluble so that it can be ejected with urine and stools (Stokowski 2011).

Recent research suggests that exposure to green light can reduce pain in people with arthritis, migraines and fibromyalgia. Green light activates photoreceptors of the retina, particularly cones, which stimulates the production of specific proteins that bind with opioid receptors in the brain and thereby diminish pain, at least in mice (Tang et al. 2022).

4.8 Healing Environments

Another use of colors in healthcare is within the environment of patients. If colors have psychological properties and influence the perception of spaciousness, coloring the interior design of hospitals and other healthcare facilities may contribute to the process of patient recovery and enhance the well-being of all users of these facilities (Fehrman and Fehrman 2018). Colors may contribute to the positive experience of these surroundings. They are first of all important for navigation, recognition and spatial orientation. Colored signage can help people to find their way within complex buildings such as modern healthcare centers. Secondly, the power of colors may be used to create healing environments within care institutions. An often-used example is exposure to green colors. This is associated with improved feelings of well-being and a more positive mood, thus bringing patients in touch with a natural and green environment may have health benefits. In a well-known study, Ulrich (1984) concluded that hospital patients in a room with a window view of a natural setting recover faster, have a shorter stay in the hospital, receive fewer negative comments from nurses, and use fewer pain medication than patients in similar rooms with a window facing a brick wall. Studies like these have stimulated research into the influences of wall color in patient rooms and the application of color theory in healthcare design. The most common color in hospitals used to be "hospital green," also called "spinach green," first invented and applied during the First World War. Before that time, hospitals and clinics were mostly painted white since that color was associated with cleanliness and purity. Also, the uniforms of healthcare workers used to be white. Surgeons, however, found white too bright, contrasting too harshly with the color of blood. It reduced their ability to discriminate anatomical features in the operating theatre. Spinach green worked much better, and brought the eyes to rest, facilitating concentration on the details of the intervention (Pantalony 2009). This is in accordance with the traditional connotations of this color: green is relaxing, calming and sanitary. Already in Roman Antiquity, green is regarded as soothing to the eyes, and it is used as eye ointment (St Clair 2016). The monotony of green in hospital settings was broken in the 1970s with the introduction of many other hues, though bright colors were mostly avoided because they were considered irritating; using only white was avoided, as was using too many different colors at once (Olgunturk et al. 2021).

The significance of color is heightened further for people living in long-term care facilities such as nursing homes. In the architectural design of such facilities, color usually is not a major consideration; interior spaces are mostly white and do not provide much sensory stimulation or convey a sense of home. It is argued that more color can enhance the quality of life in nursing homes, especially when visual health is lower, as is the case with elderly populations. Ageing often affects people's lenses, making them yellowish and opaque while the number of rods in the retina is reduced. Visual sensitivity and color perception are therefore altered, so that lighter colors and more chromatic contrast in the built environment are advisable (Delcampo-Cardia et al. 2019). Elderly residents of nursing homes themselves prefer warm hues (especially yellow) for the activity room and cool colors for the bedroom (especially green) (Torres et al.

2020). Despite numerous recommendations about the application of color in healthcare environments, the number of rigorous studies is rather limited, so that evidence-based knowledge is fragmented and uncertain. There is not sufficient evidence that particular colors in healthcare design are related to specific emotional, behavioral or healthcare outcomes for patients (Tofle et al. 2004).

4.9 Conclusion

Although color is often considered as a secondary and perhaps trivial issue in the context of healthcare, this chapter shows that it plays a role in almost all dimensions of medicine and care. Color has been a significant factor in the history of medical theory and practice, as is reflected in the naming of diseases, in the application of pigments and dyes as medication, and in the development of diagnostic tools such as uroscopy. At the same time, it was precisely the human fascination with colors that has given a significant impetus to the emergence of modern medical science and practice. If color had been considered a frivolous and insignificant issue, and had not intrigued humankind, medicine as it is known today would possibly not have developed.

The interest in colors for textiles, cosmetics and paintings has long stimulated the search for new, impressive and stable pigments and dyes. To satisfy the growing demand for colors, scientific research synthesized artificial colors in the laboratory, and developed rapidly into a major chemical industry. The new dyestuffs found specific applications within the context of medicine, particularly in diagnostics and pharmaceutical treatment. The evolution of scientific medicine on the basis of new theories and approaches in anatomy, physiology and pathology made it possible to leave the traditional humoral framework (determined by Galen) behind. This evolution was greatly facilitated because dyestuffs enabled researchers to make anatomical structures, physiological processes and pathological lesions visible and identifiable. At the time of Perkin's discovery of mauveine, Europe was affected by major epidemics. A few years earlier, England and Wales were confronted with a major outbreak of cholera, claiming many lives. Medical experts disagreed about the causes of infectious diseases. According to many, they were the result of miasmatic influences, emanations from rotten and polluted matter causing putrefied air. Others claimed that diseases such as cholera were caused by germs or contagia that were communicated between humans, and between humans and animals. The problem for this germ theory was that germs as disease-causing agents are microorganisms, not visible to the naked eye. The availability of an increasing number of artificial dyes enabled researchers to discover various species of bacteria. Since the identification in 1882 of the mycobacterium responsible for tuberculosis by Robert Koch, who used the staining method developed by Ehrlich, a host of other microorganisms have been detected within a short timeframe.

The new dyestuffs also instigated a pharmaceutical revolution, since they manifested medical properties or delivered derivatives with specific therapeutic effects. Ehrlich's observation that dyes not simply color bodily cells and tissues but produce biochemical changes and responses inspired his concept of chemotherapy, with the result that syphilis could be effectively treated with Salvarsan, the first antibacterial agent synthesized in the laboratory. Chemotherapy is now mainly used in the treatment of cancer, but for Ehrlich it was connected to the major infectious diseases of his time, and revealed the conviction that drugs targeting specific diseases could be designed and developed in the laboratory with chemical methods and instruments, devised to eliminate the causative agent. Without the persistent attraction of colors, and the possibilities of manufacturing them in chemical laboratories, contemporary medicine would not have obtained the diagnostic and therapeutic capabilities that it has today.

References

Alturkistani, H. A., Tashkandi, F. M., and Mohammedsaleh, Z. M. 2016. Histological stains: A literature review and case study. *Global Journal of Health Science* 8 (3): 72–79, https://doi.org/10.5539/gjhs.v8n3p72

Amawi, R. M., and Murdoch, M. J. 2022. Understanding color associations and their effects on expectations of drug efficacies. *Pharmacy* 10 (82): 1–23, https://doi.org/10.3390/pharmacy10040082

Anderson, M. 1990. *Colour therapy. The application of color for healing, diagnosis and well-being*. Wellingborough: The Antiquarian Press (original 1975).

Anderson, S. (ed.). 2005. *Making medicines. A brief history of pharmacy and pharmaceuticals*. London and Chicago, IL: Pharmaceutical Press.

Ball, P. 2009. *Bright earth. The invention of colour*. London: Vintage Books.

Bottaro, A. 2023. Skin discoloration: Causes and treatments. *Verywellhealth*, 12 December, https://www.verywellhealth.com/skin-discoloration-5097074

Boussel, P., Bonnemain, H., and Bové, F. J. 1983. *History of pharmacy and the pharmaceutical industry*. Paris/Lausanne: Asklepios Press.

Broadbent, A. 2019. *Philosophy of medicine*. New York: Oxford University Press.

Brock, W. H. 1993. *The chemical tree. A history of chemistry*. New York and London: W. W. Norton & Company.

Brower, J. 2022. These green books are poisonous—and one may be on a shelf near you. *National Geographic*, 28 April, https://www.nationalgeographic.com/premium/article/these-green-books-are-literally-poisonous

Buckalew, L. W., and Coffield, K. E. 1982. An investigation of drug expectancy as a function of capsule color and size and preparation form. *Journal of Clinical Psychopharmacology* 2: 245–248.

Celsus, A. C. 2021. *De medicina*. Edinburgh: University Press (original 1847), https://www.gutenberg.org/files/64207/64207-h/64207-h.htm

Chantziantoniou, N., Donnelly, A. D., Mukherjee, M., Boon, M. E., and Austin, R. M. 2017. Inception and development of the Papanicolaou stain method. *Acta Cytologica* 61: 266–280, https://doi.org/10.1159/000457827

Cleveland Clinic. 2023. Tongue color. *Cleveland Clinic*, 13 January, https://my.clevelandclinic.org/health/symptoms/24600-tongue-color

Cosans, C. E. 1998. The experimental foundations of Galen's teleology. *Studies in History and Philosophy of Science* 29 (1): 63–80.

Dathan, J. G. 1954. Acrodynia associated with excessive intake of mercury. *British Medical Journal* 1 (4856): 247–249.

De Craen, A. J. M., Roos, P. J., De Vries, A. L., and Kleijnen, J. 1996. Effect of colour of drugs: Systematic review of perceived effect of drugs and of their effectiveness. *British Medical Journal* 313 (7072): 1624–1626.

Delamare, F., and Guineau, B. 2000. *Colors. The story of dyes and pigments*. New York: Harry N. Abrams, Inc.

Delcampo-Carda, A., Torres-Barchino, A., and Serra-Lluch, J. 2019. Chromatic interior environments for the elderly: A literature review. *Color Research and Application* 44: 381–395, https://doi.org/10.1002/col.22358

De Young, M. 2015. *Encyclopedia of Asylum Therapeutics, 1750–1950s*. Jefferson, NC: McFarland & Company, 23–26.

Dyer, R. 2017. *White*. London and New York: Routledge (20th anniversary edition).

Eckstut, J., and Eckstut, A. 2013. *The secret language of color*. New York: Black Dog & Leventhal Publishers.

Fehrman, C., and Fehrman, K. 2018. *Color. The secret influence*. Solana Beach, CA: Cognella Academic Publishing (4th edition).

Gage, J. 2013. *Colour and culture. Practice and meaning from Antiquity to abstraction*. London: Thames & Hudson (original 1993).

Garfield, S. 2000. *Mauve. How one man invented a color that changed the world*. New York and London: W. W. Norton & Company.

Holzel, A., and James, T. 1952. Mercury and pink disease. *The Lancet* 259 (6705): 441–443.

Hopkins, D. R. 2002. *The greatest killer. Smallpox in history*. Chicago, IL and London: The University of Chicago Press.

Jacobs, K. W., and Nordan, F. M. 1979. Classification of placebo drugs: effect of color. *Perceptual and Motor Skills* 49 (2): 367–372.

Javaeed, A., Qamar, S., Ali, S., Mustafa, M. A. T., Nusrat, A., and Ghauri, S. K. 2021. Histological stains in the past, present, and future. *Cureus* 13 (10): e18486, https://doi.org/10.7759/cureus.18486

Jelley, J. 2017. *Traces of Vermeer*. Oxford: Oxford University Press.

Klotsche, C. 1992. *Color medicine. The secrets of color/vibrational healing*. Flagstaff, AZ: Light Technology Publishing.

Lu, D. 2020. Marie Curie's luminous legacy. *New Scientist* 245 (3274): 32–33.

Mestel, R. 2002. The colorful history of pills can fill many a tablet. *Los Angeles Times*, 25 March, http://www.latimes.com/archives/la-xpm-2002-mar-25-he-booster25-story.html

Neuburger, M. 1937. The early history of urology. *Bulletin of the Medical Library Association* 25 (3): 147–165.

Olguntürk, N., Aslanoğlu, R., and Ulusoy, B. 2021. Color in Hospitals. In: Shamey, R. (ed.), *Encyclopedia of Color Science and Technology*. Berlin and Heidelberg: Springer, 1–4, https://doi.org/10.1007/978-3-642-27851-8_449-1

Pantalony, D. 2009. The colour of medicine. *Canadian Medical Association Journal* 181 (6–7): 402–403.

Papavramidou, N., Fee, E. and Christopoulou-Aletra, H. 2007. Jaundice in the *Hippocratic Corpus*. *Journal of Gastrointestinal Surgery* 11: 1728–1731, https://doi.org/10.1007/s11605-007-0281-1

Pastoureau, M. 2010. *Les couleurs de nos souvenirs*. [The colors of our memories] Paris: Éditions du Seuil.

Pastoureau, M. 2019. *Yellow. The history of a color*. Princeton, NJ and New York: Princeton University Press.

Riesmeier, M., Keute, J., Veall, M.A. et al. 2022. Recipes of Ancient Egyptian kohls more diverse than previously thought. *Scientific Reports* 12 (5932), https://doi.org/10.1038/s41598-022-08669-0.

Riley, C. A. 1995. *Color codes. Modern theories of color in philosophy, painting and architecture, literature, music and psychology*. Hanover and London: University Press of New England.

Sallis, R. E. and Buckalew, L. W. 1984. Relation of capsule color and perceived potency. *Perceptual and Motor Skills* 58: 897–898.

Schapira, K., McCelland, H. A., Griffiths, N. R., and Newell, D. J. 1970. Study on the effects of tablet colour in the treatment of anxiety states. *British Medical Journal* 2 (5707): 446–449.

Singer, P. N. 2021. Galen. In: Zalta, E. N. (ed.), *The Stanford Encyclopedia of Philosophy* (Winter 2021 Edition), https://plato.stanford.edu/archives/win2021/entries/galen/

St Clair, K. 2016. *The secret lives of color*. London: Penguin.

Stephen, I. D., Law Smith, M. J., Stirrat, M. R., and Perrett, D. I. 2009. Facial skin coloration affects perceived health of human faces. *International Journal of Primatology* 30: 845–857.

Stokowski, L.A. 2011. Fundamentals of phototherapy for neonatal jaundice. *Advances in Neonatal Care* 11 (5 Suppl): S10–21.

Tang, Y-L., Liu, A. L., Zhou, Z. R., Cao, H., Weng, S. J., and Zhang, Y. Q. 2022. Green light analgesia in mice is mediated by visual activation of enkephalinergic neurons in the ventrolateral geniculate nucleus. *Science Translational Medicine* 14 (674): eabq6474, https://doi.org/10.1126/scitranslmed.abq6474

Time. 1956. Foreign relations: Arsenic for the Ambassador. *Time*, 23 July, https://time.com/archive/6802949/foreign-relations-arsenic-for-the-ambassador/

Tofle, R. B., Schwartz, B, Yoon, S., and Max-Royale, A. 2004. *Color in health care environments: A critical review of the literature*. San Francisco, CA: Coalition for Health Environments Research.

Torres, A., Serra, J., Llopis, J., and Delcampo, A. 2020. Color preferences cool versus warm in nursing homes depends on the expected activity for interior spaces. *Frontiers of Architectural Research* 9: 739–750, https://doi.org/10.1016/j.foar.2020.06.002

Ulrich, R. S. 1984. View through a window may influence recovery from surgery. *Science* 224 (4647): 420–421.

United Nations. 2023. Pakistan floods a 'litmus test' for climate justice says Guterres. *UN News*, 27 September, https://new.un.org/en/story/2023/09/1141587

Weisse, A. B. 1995. Tuberculosis: Why "The white plague"? *Perspectives in Biology and Medicine* 39 (1): 132–138.

Wiercioch-Kuzianik, K., and Babel, P. 2019. Color hurts. The effect of color on pain perception. *Pain Medicine* 20 (10): 1955–1962.

5. Color and Bioethics

5.1 Introduction

The world as experienced by human beings is colored. Contrary to the majority of mammals, human beings are able to see colors (as do fish, reptiles, birds and some insects). Colors make life beautiful and agreeable; they make the surrounding world pleasant and interesting. This aesthetic dimension is emphasized by Johann Wolfgang von Goethe when he states: "People experience a great delight in colour, generally. The eye requires it as much as it requires light. We have only to remember the refreshing sensation we experience, if on a cloudy day the sun illuminates a single portion of the scene before us and displays its colours" (Goethe 1970, § 759). Our delightful fascination with colors is noticeable, for example, in the sharing on social media of snapshots of skies of different colors, or in the use of impressive photos of sunrises or sunsets in television weather forecasts. Imagine what it means to live in a world without color; such a world would be dull, unappealing, uninteresting and dark. Dictionaries equate "colorless" not only with an absence of color, but with a lack of excitement or interest. In a metaphorical sense, colorless refers to a life that is sad and depressing, or to an average person without distinctive qualities. Without colors, our perception of the world would be drastically altered. As long as we are not blind, we can still see the outside world; we cannot differentiate between some colors, and in worst cases, we see only black, white and grey hues.

John Dalton (1766–1844), one of the founders of modern chemistry, discovered that his perception of colors was different from other people; he was unable to tell the difference between red and green. When studying botany, he could not distinguish flowers with certain colors, and when he bought clothes for himself and his mother which he thought were rather dark they turned out to be red (Emery 1988). His brother had the same anomaly, so Dalton concluded that it must be a hereditary disorder. He was the first to describe this condition in a presentation to the Manchester Literary and Philosophical Society in 1794. His self-diagnosed colorblindness became known as Daltonism. The inability to distinguish red and green is the most common deficiency of color vision. The most extreme, and rarest, condition is achromatopsia, i.e. a complete lack of the perception of color. Sacks and Wasserman (1987) describe the case of a painter who suddenly lost color vision after a car accident. He could only see black, white and shades of grey, and became depressed and fearful. His world changed significantly: "It was not just that colors were missing, but that what he did see had a distasteful, 'dirty' look, the whites glaring, yet discolored and off-white, the blacks cavernous—everything wrong, unnatural, stained, and impure" (Sacks and Wasserman 1987, 27). For him, the appearance of people and food was disturbing and abhorrent, and faces were difficult to identify. The world has become alien, dead and grey. In the end, he only found himself at home during the night.

The perception of color, as well as its absence, illustrates that colors have aesthetic and emotional dimensions, and a functional role in shaping our feelings. Earlier, we discussed how colors can be experienced as warm or cold. They influence our mood; when they are bright and multifarious they can make us happy; when they are absent or greyish, bleak and gloomy feelings are generated. During the Covid-19 pandemic, the best-selling paint colors were neutral, blue and green. In times of uncertainty, as one of the explanations suggests, people seek stability, comfort, healing and hope (Challener 2021). Another example of how colors may induce emotions is the recent turmoil around weather charts. Meteorologists customarily present different temperatures using colors, generally varying from blue (cold) to red (warm). Conspiracy theorists accuse meteorologists of using darker red hues to cover large expanses in order to frighten viewers and create a sense of impending doom, due to rising temperatures. According to these theories, we are not dealing with climate change but chromatic change, due to manipulation of weather maps (Nicholson 2022). This has forced weather forecast services to explain the colors of their maps (BBC 2023). On the other hand, colors are used to convey emotions. They are a means of expressing oneself, of sending visual messages through colored clothing or adornment of the body, cosmetics, tattooing and dying of facial hair. In Roman times, blue was regarded as a barbaric color since those living north of the Hadrian wall dyed their bodies blue (with woad) to appear more redoubtable in combat; they were called Picts: painted men. Since the eighteenth century, much of the indigo imported from colonial plantations was used to dye the uniforms of the police and army in Europe, while in the 1960s blue jeans became a symbol of rebellion (Balfour-Paul 1998). Displaying colors is therefore a means of communication. Depending on the circumstances and context it can be appropriate, nonconformist or wrong, indicating that using colors also may have a normative aspect, on which this chapter focuses.

Colors furthermore have a functional role. They contribute to the perception of forms and shapes; they identify boundaries between different objects and thus help to recognize objects. According to Pastoureau (2010), the first function of color is to distinguish, classify, associate, oppose and prioritize. It accentuates significant elements in the life-world of living beings. Particularly, ecological theories of color, as discussed in Chapter 2, emphasize that color vision is not simply the observation of the outside world but an instrument to identify relevant aspects of the environment which supports organisms to explore their surroundings and to survive. For many species, colors help to identify objects that are edible or toxic. They also allow organisms to adapt to their environment and make themselves relatively indetectable. The common cuttlefish, for example, is a master of camouflage; its skin has millions of chromatophores (pigments cells) enabling the animal to engage an enormous variety of skin patterns to escape detection (Woo et al. 2023). Chameleons living in the desert regulate body temperature by adapting their skin color to the weather conditions; the warmer, the whiter its skin. Chinese researchers have used this mechanism to develop a coating for buildings that changes its color depending on the outside temperature. They argue that such temperature adaptive coatings may significantly reduce energy consumption (Dong et al. 2023). Furthermore, in the animal world, colors are mechanisms of communication; they are signals to influence the behavior of other beings, as the example of the desert locusts, discussed in Chapter 2, illustrates (Cullen et al. 2022).

As argued earlier in this book, colors play similar functional roles in human societies. One function is epistemological: "colours are signs used to indicate the presence of objects of interest" (Maund 1995, 45). They enable us to identify an object, distinguish it from its background and reidentify it as the same object or an object belonging to the same class of objects. This discriminatory and identifying role of color helps a person to orientate himself in the world, and to approach some objects or persons, and to avoid others. While this role of color is apparently similar across the entire animal kingdom, human beings are unusual in using color with a normative function. It is used to articulate social divisions and to indicate social status, for example imposing yellow to stigmatize heretics, prostitutes and jews. For humans, color is not just a visual property but it is associated with a range of meanings. Yellow is a symbol of treason, deception and dishonesty, and is therefore applied to label some persons. Colors are not merely beautiful and pleasant to perceive, but at the same time function critically as symbols of good or bad.

5.2 Colors and Normativity

In Ancient Rome, purple was difficult to fabricate (made from large numbers of rare seashells) and expensive (imported from Lebanon). As a luxury color, it was reserved for high-ranking people such as magistrates and generals, and later only for the emperor (Pastoureau and Simonnet 2005). The same is true for the color yellow in ancient China. Roman writers often distinguished between somber and bright colors. The first group, *colores austeri*, are fabricated from common earth pigments (yellow, black, red and green). They are more natural and traditional. The second, *colores floridi*, are modern and exotic, commonly of oriental origin (Egyptian blue and cinnabar or scarlet). For Plinius, the first category represented the Roman ideal of *austeritas*, or severity, austerity and simplicity. The second category, by contrast, represented softness and decadence. This division was accompanied by a concern that the extravagance of bright colors would lead to an over-ornamental style, compromising the ancient ideals (Gage 2013). Roman writers such as Cicero and Seneca use the term "color" in a pejorative sense, as a figure of speech to embellish arguments. Facts are "colored" to create falsehoods and illusions (Gage 2013).

In medieval times, colors were strongly associated with symbolic meanings. Since the thirteenth century, according to Christian moral theology, the seven deadly sins have been associated with colors: envy (yellow), pride and lust (red), anger and avarice (black), sloth (white), and gluttony (green) (Pastoureau 2009, 50). Ethics therefore should not be associated with one of these colors. There are also controversies around the proper color of the religious habit (Pastoureau 2010). In the oldest monastic order, the Benedictines, the color of their clothing was initially not relevant; having a simple and inexpensive habit was most important. But over the centuries, the belief emerged that black was the most appropriate color for monks. Since the tenth century, Benedictines have been known as the black monks. For them, black is associated with humility, austerity and penitence. The order of Cistercians, separated from the Benedictines in 1098 as a movement to return to the original roots of inspiration, initially adopted grey, and later, white habits with black scapulars, arguing that white was an "angelic" color representing innocence, purity and virtue, whereas black is the color of death and sin (Pastoureau 1989).

Fig. 5.1 Cistercian monks. Bernard of Clairvaux invests Gerwig with the robes of the Cistercian order. Fresco from 1695–1698 by Johann Jakob Steinfels in Abbey church Waldsassen. Photo by Wolfgang Sauber (2018), Wikimedia, https://commons.wikimedia.org/wiki/File:Waldsassen_Stiftsbasilika_-_Fresko_3c_Gr%C3%BCndungslegende.jpg#/media/File:Waldsassen_Stiftsbasilika_-_Fresko_3c_Gr%C3%BCndungslegende.jpg, CC BY-SA 4.0.

In many societies, implicit and explicit rules commonly determine what kind of clothing people are supposed to wear (Ford 2021). There is a long tradition of so-called sumptuary laws to regulate consumption and to prevent extravagant display of luxury. An example, already mentioned, is the restricted use of purple in Ancient Rome. Such dress codes express cultural and societal norms regarding what is appropriate behavior. They also are an instrument of social control, attempting to construct social relations between people and to conserve the existing class and power structure of society. The use of colors is regulated particularly since they indicate social class and clarify the status of various groups of citizens. The Elizabethan Sumptuary Laws promulgated in England in 1574 illustrate this purpose. They state that nobody shall wear in his apparel "Any silk of the color purple, cloth of gold tissued, nor fur of sables, but only the King, Queen, King's mother, children, brethren, and sisters, uncles and aunts; and except dukes, marquises, and earls, who may wear the same in doublets, jerkins, linings of cloaks, gowns and hose; and those of the Garter, purple in mantles only" (Elizabethan Sumptuary Statutes 2001). Also the ancient regime in France used sumptuary laws to ensure the correct ordering of society through preventing lower social classes to wear certain cloths. For a long time bright colors are only reserved for the wealthy. Since black dyes are cheap, and do not adhere much to textiles, black clothes are usually worn by the lowest social classes (Pastoureau 2009). Color codes are furthermore aimed at reflecting a distinction among male and female, young and old citizens (Ford 2021). Specific colors are used to distinguish categories of people (for example, lepers, criminals and outcasts), marking them as excluded from society. Since the sixteenth century, uniforms were introduced to identify various groups of citizens and to create a sense of identity among them, such as police officers, military forces, healthcare workers, school children, lawyers and university professors.

After the Black Death in the fourteenth century, black became a fashionable color. It not only referred to death and misfortune, and the need of redemption and penance, but it also was regarded as austere and virtuous, appropriate for a particular social and professional status and a symbol of public authority. The trend towards black is already noticeable before the plague as a response to sumptuary laws introduced in 1300 (Pastoureau 2009). But the plague amplified moral concerns with color: the aspiration to restrict extravagance and to return to the tradition of temperance and virtue. This moralizing context which started in the late Middle Ages was reinforced in the fifteenth and sixteenth centuries by two developments. First, the invention of the printing press. The application of black for ink and white for paper created a "black-and-white universe" (Pastoureau 2009, 117). The second development was the Protestant Reformation, which sought to expel colors from public life, making a moral distinction between worthy and unworthy colors. The first group (white, black, brown, grey and blue) were seen as the expression of certain values such as soberness, discreteness and dignity. The second group should be avoided as disgraceful and improper. Colors like yellow and green almost disappeared from public life in some areas in Europe (Pastoureau 2019).

5.3 Color and Rationality

Moral debates about the proper use of color reflect, according to some authors, a general tendency in European culture to be suspicious of color (Batchelor 2000; Gage 1999). Already in classical Antiquity, certain colors were regarded as transgressive and morally inappropriate. They attract the eye, and capture attention, directing our mind to the surface of things rather than their essence. According to this tradition of chromophobia, colors are suspicious for multiple reasons, but an important one is that they obfuscate what is most typical for human beings. Humans are uniquely different from other living beings because they are rational animals. Rationality is characterized by discursive thinking, explanation, argument and judgment. It enables humans to distinguish what is true or false, real or questionable, good or bad. In this perspective, colors are risky since they are beautiful and appeal to the senses. They belong to the domain of emotions and subjective impressions which can vary from person to person, and from culture to culture, and therefore hinder rational analysis and objective understanding. Colors are also deceptive. It is true that everything that exists in the surrounding world used to be perceived as colored, but this hides a more fundamental reality that can only be discovered and analyzed by the mind. In fact, colors should be regarded as makeup, an envelope around objects and entities; they are merely ornamental and decorative, and should therefore be distrusted since they misguide and deceive the rational human. An illustrious debate in the history of arts focused on the question of what is more important: coloration or drawing, color or form. Opponents of color argued that design (form or line) should have priority because it is a creation of the mind; it is an expression of an idea or a concept that ultimately results in a painting, and is thus a manifestation of human intellect. Design implies a conception of the mind that is rational, structured, reliable and also honest and a sign of moral rectitude. Color, on the other hand, is emotional, rhapsodic and formless. It is not as important as the composition, subject, outline or perspective of the painting. Color may be beautiful, but it is deceitful, seducing and diverting attention from what is true and good, and even dangerous since it and its effects are not controllable (Riley 1995; Pastoureau 2010). Because coloration is dependent on the quality of pigments and materials, colors were viewed as less noble since they did not reflect the rationality characteristic to human beings.

While in the tradition of chromophobia, colors were regarded as a threat to human rationality, a normative assessment was implied as well. In medieval theology, once debate centered on whether color is matter or light (Pastoureau 2009). If color is primarily a material substance that envelopes objects, it is an artifice, a mask that conceals what is essential. This is evident from the derivation of the word 'color' from the Latin verb *celare*, which means "to hide/conceal." This is the argument of Saint Bernard of Clairvaux (1090–1153): color is opaque; it makes things dense and obscure, and does not illuminate and elucidate them. As embellishment, it is waste, a useless luxury and vanity; moreover, it is immoral in preventing humans from coming closer to God as divine light (Pastoureau 1989). The same negative attitude towards colors is noticeable in the dispute between black and white monks in the twelfth century, a disagreement which disregarded all other colors. Moral codes of color became especially endorsed and enforced by Protestant reformers in the sixteenth century, who argued, in line with the tradition of chromophobia, that color is makeup, luxury, affectation and illusion. It should be expelled from churches because the sensations of beauty and the colorful rituals and interiors distract and corrupt the sincerity of the worship of God. In painting, color asceticism should be practiced, avoiding bright colors and mainly applying black and dark tones.

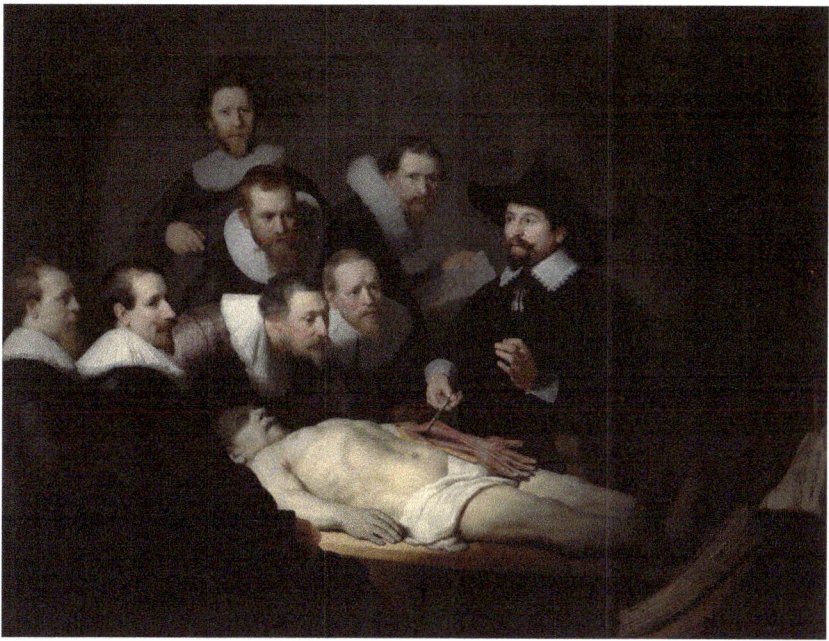

Fig. 5.2 Rembrandt, *The Anatomy Lesson of Dr. Nicolaes Tulp* (1632). Mauritshuis, The Hague. Wikimedia, https://commons.wikimedia.org/wiki/File:Rembrandt_-_The_Anatomy_Lesson_of_Dr_Nicolaes_Tulp.jpg#/media/File:Rembrandt_-_The_Anatomy_Lesson_of_Dr_Nicolaes_Tulp.jpg, public domain.

In public life, chromophobic attitudes held that people should not wear intense and brilliant colored clothing. Such moral interpretation of colors has changed human sensibilities, and also influenced the outlook of human societies at large, at least in Europe. Through advocating black and dark colors as signs of dignity, humility, austerity and simplicity "… black became the most popular color in men's clothing in Europe between the fifteenth and nineteenth centuries" (Pastoureau 2009, 132). At the same time, it became the color of mourning. The overall effect is that black and white are separated from the world of colors, and no longer considered as colors themselves. This separation is consummated in the discoveries of Isaac Newton: the chromatic sequence of the spectrum does not include black, and white is the container of all spectral colors.

The value judgments about colors in the tradition of chromophobia are regularly connected to another normative viewpoint: colors are extravagant and decadent. For a long time, numerous pigments have been imported from abroad, and this inspired the idea, already current in Roman times, that many colors have an exotic and foreign origin. This origin was used to explain why using a wide variety of colors, especially bright ones, should not be interpreted as reflecting refined taste and civilization. On the contrary, the use of bright and varied colors was thought to indicate the decline of moral values of a society, showing that traditional values such as simplicity, integrity and honesty were no longer cherished. Therefore, color was viewed as something to be purged from society because it is a property of 'foreign' bodies, a sign of otherness: oriental, primitive, infantile, female, vulgar and pathological (Batchelor 2000). It is a permanent threat because it cannot be ignored or dismissed but we have to be aware of the dangers. "People of refinement have a disinclination to colours", as Goethe writes in his *Theory of Colors* (Goethe 1970, § 841), adding that "Men in a state of nature, uncivilised nations, children, have a great fondness for colours in their utmost brightness…" (Goethe 1970, § 835). Similar ideas are expressed by the architect Le Corbusier: color is suited to simple races, peasants and savages (Batchelor 2000). For him, there is only one color: white.

5.4 Moral Associations of Black and White

That colors have a moral value is clear in the hierarchy which many societies apply to colors. Batchelor (2000) argues that cultures often oppose colors with white and black, regarded as colorless. White is associated with innocence and purity (Pastoureau and Simonnet 2005). It is a guarantee of cleanliness and hygiene. Today, many studies show that white is more highly valuated than black (Adams and Osgood 1973; Kaya and Epps 2004). White relates to goodness and what is morally preferable (Yin and Ye 2014). Black on the other hand is usually the least preferred color. It evokes negative emotions such as depression, sadness, fear and anger; it is associated with death, darkness, nighttime, mourning and tragedy; it is related to evil and immorality (Kaya and Epps 2004). Preference for white and aversion to black is found not only in Western but also Asian countries. The Chinese character for white is associated with pureness, clearness and unselfishness. In Japan, white is connected to everything clean, pure, harmonious, refreshing, beautiful, clear, gentle and natural; in Indonesia, it is associated with being clean, chaste, neutral and light, while dark tone colors are unpopular (Saito

1996). From their experiments, Sherman and Clore (2009) conclude that words with moral or immoral meanings are associated with colors. When subjects are presented with words with different moral connotations, they immediately and automatically associate immoral words (e.g. abusive, cruel, greed, hate and revenge) with the color black. Moral words such as duty, freedom, honesty and justice, on the other hand, activate the color white. But why is black connected to evil and immorality? Sherman and Clore explain the connection with the thesis that physical purity is a symbol for moral purity. If moral goodness is associated with physical cleanliness, and thus white, the implications are particularly negative for black. It is not just the opposite of white, but it may contaminate and pollute white, make it dirty and impure. If white represents morality and virtue, black stains and perverts it and introduces immorality (Sherman and Clore 2009).

The idea that blackness has polluting powers and is associated with sin and moral evil is derived from anthropological theories about the notions of purity and pollution. Most human societies are concerned with preserving things in an original and uncompounded state, and have rituals and practices of cleanliness and purification. Systems of classification separate practices and activities that are considered valuable from those that are dirty and impure, and should be averted (Forth 2018). It is interesting that in Ancient Rome, writers arguing that basic colors should express traditional simplicity also articulate that they should not be mixed since that produces change, putrefaction and conflict. In the early nineteenth century, it was commonly thought that classical Greek marble sculpture and architecture was, by design, pure white; research showing that statues and temples were traditionally colored came as a shock to Victorian culture (Gage 2013). Around the same time, concerns about contamination increased with the production of artificial dyes. The use of natural pigments and dyes commonly delivered a product that was not completely reliable, stable and durable since they often contained other impure materials. Production of such dyes also led to significant pollution of rivers and environments. Chemical fabrication, however, aims at a constant and predictable color, so it entails a vast effort at purification, eliminating dirt and traces of natural pigments (Brusatin 1986). This reflects the contemporary ideals of hygiene and cleanliness. The nineteenth century was the age of the

Industrial Revolution, causing a blackening of the environment with smoke, coal, tar and soot, but also through social repercussions like overcrowding of cities, poverty, child labor and epidemic diseases such as cholera (Harvey 2013). The sanitary movement initiated a struggle against dirtiness, trying to control communicable diseases that ravaged, in particular, urban areas, through programs to remove waste, reduce water and air pollution, improve sewage systems and generally clean up the environment. Filth was regarded as the cause of disease, and as the mode of disease transmission. From a hygienic perspective, all things black should be avoided since they were thought to be harmful, dangerous and contaminating. Around this time, physicians, who used to dress in black, started to wear the modern white coat (Seeman 2017).

Anthropological theories of purity and pollution do not fully explain why black and white are associated respectively with negative and positive emotions, and thus regarded as impure or pure. Another perspective emphasizes the importance of experiences. Already early in life, human beings go through the alternation of day and night, light and dark. Since humans are typically diurnal, we tend to be active during daytime: for activities, we need light, and when light is diminishing or absent, we become less active, and rest or sleep. This circadian rhythm explains the preference for light over darkness. It may also clarify why preschool children have an aversion to darkness, and may experience disorientation, fear and deprivation in the dark (Boswell and Williams 1975). At the same time, there is also a history of cultural experiences. Black is regarded as a primordial color, because it is one of the oldest pigments used in paintings (for example, in Paleolithic caves), but also because of its role in mythological and religious creation stories that generally assume initial darkness and blackness. In Western Antiquity, colors are connected to the four basic elements of the physical world. Galen, for example, relates black to earth. In many cultures, black is the color of death. In Ancient Egypt, Anubis, the god of death is represented as a black jackal. In Ancient Greek times, the subterranean world ruled by the black god Hades is all black. In Christianity, hell and the devil are imagined as black (Pastoureau and Simonnet 2005). Darkness is furthermore equated with sin. While sin is traditionally conceived as red (as the color of blood), it is regarded as a stain upon the soul, darkening the light of God, making black into a sign of evil (Harvey 2013).

These cultural connotations of black refer to another interpretation of the negative associations of this color (Kareklas, Brunel and Coulter 2014). Growing up and learning to adapt to a cultural setting means internalizing and comprehending the color symbolism of that setting. Nearly all cultures attribute negative qualities to black: it refers to death, depression, tragedy, misfortune, terror and negation, and also to evil and wickedness. But even so, black has an ambivalent meaning, since it additionally has positive attributes such as humility and penitence (as shown in the discussion of the monk's habit), and authority, professional expertise, seriousness and distinction. Apparently, cultural contexts first of all articulate the pejorative associations of black, as is reflected in numerous negative expressions in everyday language (e.g. black day, black market, blackmail, blackout, black hole and black sheep) (Frank and Gilovich 1988). As humans are educated and acculturated, they learn to develop automatically preferences for white and aversions to black. Anthropological, biological and cultural theories provide different explanations of why the color black has negative associations, but they all lead to the same result: it is a symbol of evilness.

5.5 Color and Race

As discussed earlier, one of the major functions of color is to make distinctions: to identify objects and entities and to differentiate among them, connect or oppose them with each other, and classify them. This functional role is linked to emotions and normative associations, interpreting some colors as good or desirable and others as bad or unwelcome. The moral value attributed to color is evident in its use to articulate social divisions and distinctions. For example, social classes have historically been indicated by the colors that they are allowed to use for their clothing. These functions and associations of color become problematic and disputable when color is connected to race.

In the seventeenth century, the concept of race begins to emerge as a way to divide the human species into distinct groups on the basis of biological differences. These differences are manifested in physical phenotypes, and skin color is one of the most visible characteristics. The French physician Francois Bernier (1625–1688) was one of the first to present a racial classification in 1684 (Stuurman 2000).

Fig. 5.3 Jean-Auguste-Dominique Ingres, *Portrait of Francois Bernier* (1800). Wikimedia, https://commons.wikimedia.org/wiki/File:Bernier-Ingres-1800.jpg#/media/File:Bernier-Ingres-1800.jpg, public domain.

He argued that rather than classifying human beings on the basis of geography, physical characteristics as objective criteria should be used, such as skin color, facial type and bodily shape. Bernier distinguished four species or races. The "first race" was defined by whiteness, and included Europe, North Africa, the Middle East and India as well as the native population of the Americas and some parts of South-East Asia. The "second race" consisted of sub-Saharan Africans, with blackness deemed an essential trait by Bernier, who also associated the grouping with "savagery."

The emergence of racial classification in the late seventeenth century can be variously explained with references to the prevailing context. Due to colonial expansion, interest in travel literature and ethnographic descriptions such as those provided by Bernier, who lived in India for many years, intensified (Stuurman 2000). Exploration of new areas of the world, of different cultures and societies engendered confrontation with what was "foreign" and "other." At the same time, it produced a deluge of new knowledge of objects, ideas, customs and languages that needed to be understood and explained. The rising influence of empiricist philosophy (e.g. John Locke and Pierre Gassendi) encouraged the interest in natural history and taxonomy (Hannaford 1996). It also stimulated empirical approaches in describing and analyzing differences and inequalities, as well as efforts to reduce the multitude of particulars to general categories.

Initially, the status of color in classifying human beings was unclear and muddled. This is evidenced in the work of Bernier, whose "first race" covers not only Europe. Native Americans, for example, are included in his grouping, although they are "olive-colored." Bernier also viewed Chinese and Japanese people as "really white," but based on other physical characteristics he assigned them to a separate class (Stuurman 2000). Bernier's concept of race is thus a curious construct; nonetheless, his classification importantly introduced categorization of humans according to physical characteristics, taking whiteness as a point of comparison and opposition for "others." In the eighteenth century, more systematic schemes were produced in which color is attributed a decisive role. In 1735, Carolus Linnaeus classified thousands of species of animals and plants. He divided the human species (*homo sapiens* in his terminology) into four "varieties": americanus (reddish), europaeus (whitish), asiaticus (tawny) and africanus (blackish). His observation of different colors aligned with racist judgments of character traits: Europeans were not only white but also serious, strong, active, smart and inventive, while Africans were black, impassive, lazy, slow and foolish. Linnaeus' taxonomy is considered the prototype of scientific classification and it inspired numerous racial classifications through introducing a polarity between white Europeans and black Africans (Stuurman 2017). Major contributions to the theory of human diversity were made by George-Louis Leclerc, Comte de Buffon and Johann Friedrich Blumenbach. For Buffon (1707–1788), skin color was the main feature and marker of the four human "varieties," although the varieties all ultimately represented the same human species (Eze 1997). Different colors, in his view, were the result of ecological factors, especially climate (exposure to sunlight), food and way of life: "Man, white in Europe, black in Africa, yellow in Asia, and red in America, is always the same man, taking his color from the climate" (Buffon, in Stuurman 2017, 303). However, for Buffon, white is the true color, the global standard; nonwhites are degenerated from this original, and thus inferior. The German physician and anthropologist Blumenbach (1752–1840) distinguished five varieties of the human species ("races") according to skin color: Ethiopian (black), Caucasian (white), Mongolian (yellow), Malaysian (brown) and Amerindian (red) (Eze 1997).

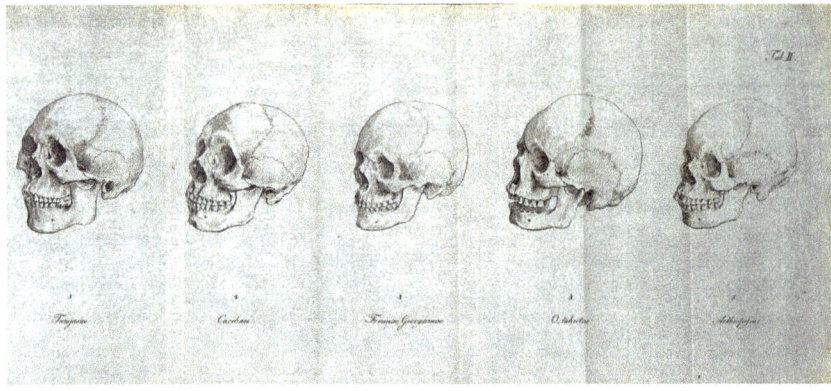

Fig. 5.4 Johann Friedrich Blumenbach, *De generis humani varietate* (1795). Sequence of human skulls showing the diversity of the main types. Wellcome Collection. Wikimedia, https://commons.wikimedia.org/wiki/File:J.F._Blumenbach,_De_generis_humani_varietate_Wellcome_L0032295.jpg#/media/File:J.F._Blumenbach,_De_generis_humani_varietate_Wellcome_L0032295.jpg, CC BY 4.0.

He argues that the differences between these varieties are so small and gradual that it is almost impossible to make sharp distinctions. At the same time, he strongly opposes any hierarchy among the varieties, rejecting the ideas that some are superior and others inferior (Pastoureau 2019). Nonetheless, he views skin color as "the most constant of all bodily differences between races" and "white is the natural complexion of humanity" (Stuurman 2017, 310, 311). Buffon and Blumenbach point out that "races" are not natural kinds but the product of environmental circumstances and that the boundaries between "races" are not fixed but dynamic and arbitrary. Assuming the reality of racial categories in their taxonomies, however, they inspired the development of racial theories (Malik 2023).

5.6 Skin Color

Examining the origins of racial classifications demonstrates how color comes to be associated with race. The application of the term "race" to human being is relatively new: contrary to what modern racist theories have suggested, the term was not used in ancient and medieval times. In Western languages, it came only into general use in the middle of the sixteenth century (Hannaford 1996). Human differences were recognized and described but not conceptualized in terms of race. They were commonly attributed to place and geography, and explained with references to climate which was thought to produce different constitutions, temperaments and characters. A distinction was made not between colored and non-colored but rather between "civilized" and "barbaric" people (Painter 2010). The latter category was considered primitive, savage and alien; their perceived inferior nature was used to justify enslaving "savage" people. In Antiquity, slavery was a widespread practice and most enslaved people were white (in Ancient Greece, they came from the Black Sea region; in Ancient Rome, they were Gauls, Germani and Celts).

The focus on skin color as determinative of race is also new. For a long time, the concept of "skin color" had no useful meaning (Painter 2010). That does not imply that no distinctions between people were made, but they were based on other criteria such as rationality and civilization. In ancient thought, color itself was regarded as a kind of skin, as a surface rather than substance, susceptible to change and movement. Within the humoral framework of Hippocrates and Galen, the colors of the skin reflected the balance or imbalance of the bodily fluids, and were thus helpful in diagnosing health or disease. In this theoretical framework, calling a person 'white' meant that he or she was anemic or even moribund. A particular coloration of the skin therefore was not a marker of a specific human species. The change in meaning of 'skin color' only became possible when humoral theories gave way to new physiological and pathological theories. Nevertheless, some elements of the old framework persisted in eighteenth-century classifications. Linnaeus characterized the American as choleric, the European as sanguine, the Asian as melancholic and the African as phlegmatic (Eze 1997). This reflects the ancient idea that the humors were associated with different temperaments or personalities, though it does not apply the traditional theoretical ordering of colors (for example, black bile used to be associated with a melancholic temperament, and a phlegmatic temperament with white). The question remains why the new idea of "race" came to be connected with color. Why is skin color regarded as the "keystone trait" to classify people? (Jablonski 2021). An obvious explanation assumes that color is a directly and publicly visible quality, making race into an evident visual experience that is "objectively" observable. Since it is inherent in the human body, it can be taken as a natural phenomenon that is independent from the context of the observer. Color is a physical attribute that immediately signifies human difference. However, it is remarkable that classifications commonly distinguish only four colors. In Western culture, since the Middle Ages a chromatic system with six colors has prevailed (Pastoureau 2001). Before this, four basic colors were identified (black, white, red and yellow), as in Galen's system that dominated medical thinking since Antiquity, relating the macro-cosmos to the micro-cosmos (earth and black bile/black; water and phlegm/white; fire and yellow bile/yellow, and air and blood/red) (Hoeppe 2007). For a long time, these colors were used to distinguish the four stages of alchemy. Although the cultural color system has changed, and Newton even identified seven spectral colors, racial classifications continue to use the classical system. Green and blue were not selected in association with race; perhaps these colors evidently indicated morbid conditions. The four-color scheme furthermore ignored existing diversity. Bernier included native Americans in his "first race" although they were not white, while Buffon noticed that Africa is "remarkable for the variety of men it contains" (Buffon, in Eze 1997, 20). He also makes a distinction between two kinds of black people: "Negroes," the blackest men in the western territories, and "Caffres," men of less deep blackness on the eastern coasts (Eze 1997, 22). Blumenbach acknowledged that the boundaries between his five "varieties" were not clearly demarcated. As skin color is the result of ecological factors it varies according to the heat of the climate, allowing for gradual variation of skin colors. Even in Europe, White people are not or not all white, as Buffon remarks—the burning sun "makes the Spaniards browner than the French" (Buffon, in Stuurman 2017, 304).

Another sign that race is not equivalent to actual skin color is that people are included in racial categories who evidently do not have the color of that category. A well-known example are Irish immigrants in the United States. In the nineteenth century, they were judged as racially different from Anglo-Saxon Americans, and put in the same category as Black people. Discrimination and marginalization were aimed at two inferior races (Celt and African) (Painter 2010). That the notion of race has a political and social function without references to skin color is Theodore Allen's thesis (2021). He argues that "race" and the privileges attached to whiteness have been used by ruling classes to maintain social control and to justify oppression and slavery, comparing Irish and United States history and identifying analogous mechanisms of racial control and exploitation (Allen 2021). These examples illustrate that what is determinative for a specific race is not the actual color of the skin but rather its associated normative connotations, the idea of color. It provides another explanation of the use of skin color in racial classifications: color functions as a code for moral worth and character. The idea of whiteness became representative of rationality, freedom, morality, and beauty, whereas the idea of blackness came to evoke irrationality, primitivism, ugliness and a slavish nature. Nell Irvin Painter (2010) clarifies this point by demonstrating the enlargement of the concept of whiteness in United States history. Initially, only Anglo-Saxon males counted as "American"; later the Irish and Germans were included, then the Southern and Eastern Europeans, and still later Asians and Latinos (for instance, in the 1930s, federal and Texas state law defined Mexicans as white). What persisted was the dichotomy between black and white; black continued to be connected to evil and negativity, while "white" people were believed to be superior (Jablonski 2012).

5.7 Color-Based Hierarchies

The seventeenth-century taxonomies of Linnaeus, Buffon and Blumenbach used skin color as the primary criterion to classify people into different groups. For describing and understanding the natural world, this is considered as a self-evident criterion. Taking color as classificatory criterion, however, transforms it from an accidental observation into a signifier of human difference. For Bernier, for example, Egyptians and Indians were very black or copper-colored, but that color is only accidental (due to exposure to the sun) whereas the blackness of sub-Saharan Africans was not caused by the sun but due to "the peculiar texture of their bodies, or… the seed, or --- the blood" (Bernier, in Bernasconi and Lott 2000, 2). While previously, skin color was merely related to geographical origin and location (where it could vary according to temperature and climate), it now became a marker of "real" differences between human beings (Jablonski 2021). As a natural phenomenon and physical trait, it was no longer regarded as the effect of external conditions in which people lived but interpreted as the reflection of an inner state, as an inherent characteristic of the body (as in the view of Bernier) and perhaps also of the mind or personality. Linnaeus connected colored races to specific characters: Whites were gentle and inventive, and governed by laws, whereas Blacks were indolent, negligent and governed by caprice (Eze 1997, 13). Immanuel Kant posits that the difference between Black and White races "appears to be as great in regard to mental capacities as in color" (Kant, in Eze 1997, 55). In his opinion, "Humanity is at its greatest perfection in the race of the whites" (Kant, in Eze 1997, 63): White people excel because they have a beautiful body, work harder, control their passions, and are more intelligent than other races. That Black and White people differ not only physically but also psychologically was, furthermore, argued by Thomas Jefferson, who thought that the existence of the first group was more determined by sensation than reflection. In his view, the distinctions between the two races were produced by nature, not by the conditions of life (Eze 1997, 98–99).

These judgments about the physical and mental qualities of races make clear that classifying humans on the basis of color implies a ranking and hierarchy of races; at the same time, the suggestion is that the division of humankind is based on scientific, "objective" criteria. Using only physical traits for their classifications, Blumenbach and Buffon regarded white as the standard from which other colors have "degenerated." For Blumenbach, White is the most beautiful race; for Buffon, it is the genuine color of mankind. When races are also connected to mental characteristics, the ranking becomes even more prejudiced: some races were viewed as clearly superior, and others inferior. Non-

White races have innate inferiority, according to David Hume, and for Jefferson, Black people are "inferior to the whites in the endowments both of body and mind" (Jefferson, in Eze 1997, 102). The normative associations of colors that have existed since Antiquity before the concept of race was invented, especially those related to black and white, were projected on human beings themselves. In the new idea of races, white, with its mainly positive connotations, was taken as the primary point of departure to produce contrasts with other skin colors with mostly negative associations. This projection can be aesthetically motivated, as when Blumenbach emphasizes the beauty of the White race. This is in accordance with the ideas of Johann Winckelmann (1717–1768), the influential art historian, who argued that white Ancient Greek and Roman sculptures represented perfect human beauty; color in statues meant barbarism (Painter 2010). For many others, the projection of colors implies a moral difference which is, in fact, reminiscent of much older ideas that color is not rational, not a manifestation of the human intellect, as well as at the same time being foreign and primitive.

5.8 Racial Science

The creation of races on the basis of color has set the tone for subsequent racial science in the nineteenth and twentieth centuries. Skin color continues to be a marker of racial difference, and is sometimes used as synonym of race, but in the words of Jablonski (2021, 442), "it no longer took center stage." Other markers of race classification that had the allure of objectivity became important, and were assumed to be measurable and quantifiable, such as cranial shape and size, genetic constitution and intelligence testing. The development of racial science has been extensively examined and criticized, and it will not be elaborated here (Hannaford 1996; Valls 2005; Painter 2010; Saini 2019; Zack 2023; Smedley et al. 2024). Nevertheless, in the context of bioethics, two observations are important. The first is that medical doctors significantly contributed to this development. Blumenbach (in Göttingen) had a large collection of skulls, and promoted craniometry as an objective, measurable parameter of race. Samuel Morton (1799–1851; in Philadelphia) was an authority in the physical measurement of skulls, using the volume of the cranium to determine brain size. From his "empirical" data, he concluded that races could be ranked according to the average sizes of their brains, and that innate differences in intellectual capacity could be measured. Later re-examination of his data showed that there are no significant differences among races, and that in fact a prior racial prejudice had determined the ranking and interpretation of empirical findings. Morton obtained the results that he expected (Gould 1996). A similar conclusion is drawn for the work of Paul Broca (1824–1880; in Paris) who concluded that his study of brains and skulls demonstrated that the development of intelligence related to the volume of the brain, and that this finding was evidence for a hierarchical ranking of races. Quantification and rigorous science led to the conclusion that the brain is larger in superior races with a white skin. But critical review of his work reveals that quantification was used to illustrate a priori conclusions (Gould 1996, 114 ff). An influential promotor of scientific racism was Louis Agassiz (1807–1873) at Harvard. He underlines that races are separate species with different innate value; they do not have the same abilities, dispositions and powers. Because of these natural inequalities, races should be treated differently and they should be strictly separated (Gould 1996, 74 ff). A last example of the contribution of medical doctors to race science, using comparative anatomy in particular to show the inferiority of some races, is Cesare Lombroso (1835–1909; in Turin). He identified anatomical characteristics ("stigmata") of criminality, comparing criminals with inferior races. One of his stigmata was darker skin (Gould 1996, 159).

Another observation concerns the contribution of philosophers to the development of racial science. Immanuel Kant is regarded as the first to elaborate a theory of race (Bernasconi and Lott 2000; James and Burgos 2023): the White race and the Negro race are the basic races, and the reason according to Kant is self-evident. This is not further explained but presumably refers to color, among other traits, since the Hindustani race is characterized by olive-yellow skin color, and the Kalmuck race by red-brown color. For Kant, skin color is the most important characteristic which is hereditary, and which determines the difference of races (Sandford 2022). The four human races that Kant identifies originate from "germs" or "seeds" (*Keime*) and "natural dispositions" (*natürliche Anlagen*) which determine the development of the organism. Though this stem genus is now extinct, Kant asserted that white inhabitants of

Northern Europe were closest to this original form. Races as permanent features determine the hereditary character of peoples. In Kant's view, skin color and character are directly connected. Color is evidence of moral quality. While Europeans can progress in the perfection of human nature, and are thus able to improve themselves, other races are incapable of moral advancement. The question of how the acceptance of racial views affects and compromises Kant's philosophical and ethical theories is the subject of intense recent debate. Is his concept of personhood (with rationality and capacity for autonomy as distinctive) and his theory of moral agency regarded as a universal characteristic of humanity, or is Whiteness a condition meaning that the concept of humanity cannot be extended to other races (Mills 2005; Marwah 2022)?

That medical doctors and philosophers have contributed to the establishment and development of racial science is not a coincidence. As a modern invention, emerging in the Enlightenment, the concept of race reflects "a new ordering of things according to nature" (Hannaford, 1996, 154). It is based on the belief that rational science (particularly physical anthropology and comparative anatomy) can explain differences between humans on the basis of structural (physical and anatomical) characteristics (interpreted as observable "facts") rather than traditional references to varying political, social and religious settings of life. In the nineteenth century, medicine became dominated by biological determinism. The ecological approach of Blumenbach and Buffon was rejected and replaced with the belief that human nature is determined by intrinsic and unalterable physical, chemical and biological constituents which can be measured and quantified. Emphasis shifted from nurture to nature, first by studying the phenotype (with craniometry and biometrics), followed by increasing insistence on genetic determinants and genotype, but still assuming race to be a natural attribute of human beings. Scientific research on these biological and genetic determinants pretended to offer an objective approach, but in fact implied a subjective ranking of human beings. Racial examination and classification always entails a hierarchical ordering (Stuurman 2017, 344). Identification of separate races leads to the conclusion that they are unequal. Since inequality is considered as a biological fact, and white skin is regarded as the standard, distinction of races has particular consequences. First used to justify discrimination, segregation, slavery and colonialism, it is later used to advocate restrictions on immigration, intermarriage and compulsory sterilization in order to counter racial mixing, and to criticize social programs and services because biology was understood as unchangeable. Gould (1996) blames Blumenbach for initiating this shift in perspective; he is the first to introduce a change from a geographic ordering of human diversity towards a hierarchical one. Color no longer refers to environment and geographic location but to biological, cultural and behavioral differences (Jablonski 2021). Taxonomies therefore are a specific manifestation of one of the traditional functions of color: the urge to classify, distinguish and separate (Saini 2019). Using the color of the skin as a hallmark of human races serves to justify different treatment, particularly when physical characteristics are connected to mental capacities, morality and character, as argued by Enlightenment philosophers. The influential work of Arthur de Gobineau (1816–1882) exposes the implications of taxonomic approaches; since all civilizations derive from the White race, and races are unequal, mixing produces decadence and decline of civilization (Malik 2023; Smedley et al. 2024).

5.9 The Persistence of Race and Racism

Nowadays, the scientific consensus is that there is no evidence that the cultural classification of "race" corresponds to an underlying biological or genetic reality. Races are cultural and social inventions used for political and ideological purposes (Smedley et al. 2024). The concept of race is a fiction; it does not correspond to an objective reality in nature, and it is therefore erroneous and meaningless (Montagu 1941). However, the concept has not disappeared from public discourse. The UNESCO statement on race strongly argues that there are no differences in innate intellectual and emotional capacity between people, and that inherited differences are not a major factor in producing cultural differences and achievements, but it still assumes that races do actually exist stating that the use of the word "race" should only be limited to groups of humans that have "well-developed and primarily heritable physical differences" (UNESCO 1952, 11). An oppositional view argues that the word "race" is better eliminated from our vocabularies. As long as the word continues to be used, it will be difficult to avoid its negative connotations and discriminatory responses. Blum (2002) proposes discarding the term "race" (and other racial words) and to use "racialized groups" instead. This allows us to retain ability to identify and criticize racism because it acknowledges that groups are treated and regarded as a race, with characteristics that are negative, inherent and immutable. It also indicates that races are not simply social constructs; they do not exist whereas racialized groups are real, as social creations.

Nonetheless, racial thinking and language have not disappeared from contemporary societies. Racist scientists, networks and journals remain active, even today (Wilson 2024). Gould (1996) and Saini (2019) give many examples of racial theory in the last few decades, leading Gould to conclude that "the same bad arguments recur every few years with a predictable and depressing regularity" (Gould 1996, 27). The concept of race, and particularly various skin colors, continues to be used to justify distinct treatment of individuals and groups (Omi and Winant 2000). Although there are differences between contemporary and classic notions of race, as argued by Blum (2002, 132), explicit racism has diminished, and racial discrimination is legally prohibited in most countries, racial thinking and racist practices are still present. In February 2021, the UN High Commissioner for Human Rights concludes that "Racism and racial discrimination occur daily to millions of people around the world" (United Nations 2021). Racism and racial incidents are reported in a wide variety of countries. For example, in Germany twenty-two percent of the population indicate that they have been victims of racism (DeZIM 2022). A recent survey in the Netherlands reveals that ten percent of government officials experienced racism in the workplace, and that eleven percent observed racism by colleagues towards citizens, despite the official policy of equal treatment, diversity and inclusion (Rijksoverheid 2024). While numerous studies of racism have been done in the United States, racial discrimination is not less of a problem in many other countries. For example, in the labor market, racial discrimination in hiring is higher in France and Sweden than in the United States (Quillian et al. 2019).

Whether or not the concept of race is applied or distinctions between races are deemed meaningful, the general consensus is that racism is morally objectionable. "Racism" is a relatively new term, used first in the beginning of the twentieth century. It articulates the deleterious consequences of the notion of race: discrimination, exploitation and denial of dignity (Blum 2002). The consensus that these consequences are unacceptable is expressed by the international community in the *International Convention on the Elimination of All Forms of Racial Discrimination*, adopted in 1965 by the General Assembly of the United Nations (United Nations 1965) and entered into force in 1969. Its purpose is to eliminate all forms of racial discrimination, i.e. "any distinction, exclusion, restriction or preference based on race, colour, descent, or national or ethnic origin". Although the notions of race and racism are differentiated, racism is sometimes defined in a way that is hardly distinguishable from the theory of race, for example as "the belief that humans may be divided into separate and exclusive biological entities called 'races'" (Smedley 2024). It is evident that racism presupposes the belief that races actually exist (with all implications of separation, segregation and hierarchy, as discussed previously), but it is more than a system of beliefs, ideology or theory: it also implies behavior, attitudes and social practices towards specific race-defined groups. Racism, according to Blum is defined by "inferiorization" (as mostly implied in racial theories) and "antipathy," i.e. disrespect and contempt, hostility and hatred, manifested in "actions, motives, attitudes, statements, symbols, images, practices, societies, and persons" (Blum 2002, 5, 8). Racism has notable consequences: it "creates or reproduces structures of domination based on essentialist categories of race" (Omi and Winant 2000, 206). It is not simply a manifestation of individual prejudices, but is also expressed in sociocultural and institutional arrangements.

5.10 Racism and Healthcare

Recently, prestigious scientific journals such as *Nature* and the *New England Journal of Medicine* have acknowledged their complicity in race theory, racism and slavery (Editorial 2020; Jones et al. 2023). They have disseminated racial views and have justified racism under the guise of scientific evidence and theory. But they also ascertain that this is not history: the scientific enterprise remains complicit in systemic racism (Nobles et al. 2022). While support for racist policies as well as explicit racism have significantly declined over the past few decades, there is substantial evidence that racist beliefs, attitudes and behaviors play a role in healthcare practices. Patients from racialized minorities report overt and covert racism in interactions with healthcare providers. They experience lack of involvement in decision-making processes, lack of respect, rude treatment, negative stereotyping and feel that symptoms and complaints are not seriously considered (Hamed et al. 2022). Physicians describe how they encountered racist behaviors in their training and practice (Tweedy 2016; Calhoun 2021; Blackstock 2024). Studies from a range of countries show that racialized minority healthcare staff experience racist behavior from other healthcare providers, as well as patients. They are expected to tolerate such behavior from healthcare users because the latter are sick and vulnerable (Hamed et al. 2022). In the UK National Health Service, 29.8% of healthcare workers from the Black, Asian and minority ethnic community experienced bullying or abuse from patients or the public in 2018 (compared to 27.8% of White NHS staff). They were afraid to express concerns within an organizational culture which does not formally regard race as a relevant issue (Danso and Danso 2021). Numerous studies describe racist beliefs and attitudes of healthcare providers: they tend to regard patients from racialized minorities as less reliable and cooperative, more problematic, frustrating, irrational, too demanding and too emotional (Hamed et al. 2022; Ray 2023).

Evidence of racism in healthcare is frequently explained as the result of implicit bias, as produced through unconscious and involuntary processes (Hall et al. 2015; Williams and Wyatt 2015). For example, in the United States healthcare providers, like the general population, "have implicit biases against Black, Hispanic, American-Indian and dark-skinned individuals" (Maina et al. 2018). These biases, especially negative attitudes towards people of color, lead to poorer interactions with these patients as well as poor health outcomes. Empirical research shows that racial prejudice influences medical decision-making and treatment decisions (Paradies et al. 2014). An often-mentioned example is that referral rates for specialist services, and prescriptions of pain medication, are lower for Black patients compared to White patients (Hamed et al. 2022; Ray 2023). The biases of healthcare professionals often result in differential treatment; consequently, healthcare users experiencing racism not only distrust healthcare but also avoid seeking care, and have lower medication adherence. The experience of racism and discrimination furthermore is a psychosocial stressor with negative effects on health, and especially mental well-being (Williams and Mohammed 2009; Lewis et al. 2015; Paradies et al. 2015). Exposure to discrimination, for example, increases the risk of psychotic disorders in members of ethnic minority groups (Veling et al. 2007).

Explaining racism and racial discrimination in terms of implicit bias reflects the current situation in which explicit expressions of racism and overt practices of discrimination are no longer accepted. Such expressions and practices are directly recognizable, and they have obviously decreased due to prevailing social, legal and ethical norms in most societies. However, racist beliefs, attitudes and behaviors have not dissipated but have become less visible and noticeable, and more subtle because they are often unconscious, unintentional and even involuntary. This explains why racist practices persist despite formal rejection by most people. Nevertheless, in the context of healthcare, the psychological explanation faces two difficulties. First, it is against the prevailing medical morality. Healthcare ethics emphasizes impartiality, neutrality, objectivity and thus the significance of equal and just treatment. Differential treatment generated by racist prejudice goes against the core ethical principles underlying healthcare and medicine. Most healthcare practitioners therefore tend to dismiss racism and deny that it exists in healthcare. Usually, experiences of racism are not discussed in the workplace (Hamed et al. 2022). The second difficulty is that implicit bias as psychological explanation of racism focuses on beliefs, attitudes and behaviors of individuals. Even when it is admitted that such biases exist within healthcare, they are regarded as exceptional; some individual healthcare providers obviously do not implement the generally accepted ethical principles underlying healthcare. These biases should be made conscious and explicit, and the involved persons should be retrained and better educated in healthcare ethics. The problem with this individual focus is that it presents only one level at which racism can occur. Racism is also manifested in institutional structures and policies.

In the recent literature, a distinction is made between interpersonal racism (at the level of interactions between individuals) and institutional or systemic racism (at the level of policies, practices or processes within institutions and organizations). The second level of racism is considered as the most fundamental one (Jones 2000). It offers a social and political, rather than psychological, explanation since it interprets racism as a system of social inequality: "Racism is about power and dominance; about ethnic and racial inequality, and hence about groups and institutions and more complex social arrangements of contemporary societies"

(Van Dijk 1999). This level is also the most pernicious one for health since it produces systemic and structural disadvantages for racialized groups rather than merely individuals. The disadvantages and resulting inequities are also not the outcome of individual agency. The oppressive and discriminating mechanisms are less identifiable than racial acts of individuals (Elias and Paradies 2021). Systemic racism is, for example, reflected in reduced accessibility to goods, resources and services, such as education, employment and health insurance. Because of residential segregation, racialized minorities have poorer health services, unhealthy living conditions and exposure to toxic environments (Paradies 2017).

The existence of health disparities is widely recognized today. The WHO Commission on Social Determinants of Health concludes in 2008 that human health is determined more by the conditions under which daily life is lived than by medical treatment and healthcare services (WHO 2008). Previously, the US Institute of Medicine report *Unequal Treatment* stated that race and ethnicity are significant predictors of the quality of healthcare received, even after differences in socio-economic conditions are accounted for (Institute of Medicine 2003). As argued in the discussion above, inequities in health and healthcare, and specifically racial and ethnic disparities, persist despite these reports. The question is how these disparities can be explained. A survey among doctors and nurses shows that they prefer individualistic explanations, such as the belief that Black patients have lack of compliance, are not well informed, take less control over their care, miss appointments and are hesitant to accept referrals to specialists. Provider bias is also regarded as a reason for unequal treatment. The most commonly mentioned systemic factor was lack of insurance coverage. Incidentally, half of the respondents questioned the validity of research studies documenting racial disparities (Clark-Hitt et al. 2010). Explanations of health disparities on the basis of individual characteristics of healthcare users or providers are insufficient since they cannot clarify why inequities are so pervasive, widespread and enduring except by referring to individual prejudices. Another type of explanation emphasizes the socio-economic context of health and healthcare. If the conditions under which people are born, grow, live, work and age have significant impact on health, it should be acknowledged that for racialized minorities those conditions are generally worse than for non-racialized majorities. Health disparities should therefore be explored at the structural level of the social, environmental and economic contexts in which people are embedded. However, this type of explanation raises the question of how differences in socio-economic status can be explained. Here a third type of explanation is introduced, referring to the long history of racism and racial exploitation, at least in the United States. Racial and socio-economic status are "closely interwoven" (Feagin and Bennefield 2014, 8). Structural explanations of disparities are important because they go beyond the idea of implicit bias, but they "do not offer a sufficient explanation for persistent racial differentials" (Feagin and Bennefield 2014, 12). At the heart of health inequalities is not merely bias but systemic racism which has produced unequal socio-economic conditions due to the accumulation of resources by generations of White people who have benefitted from slavery and racial oppression and have denied such resources to people of color. The legacy of racial discrimination and exploitation which have been pervasive for a long time is still felt today.

Fig. 5.5 "Colored" water cooler in streetcar terminal in Oklahoma City (1939). Wikimedia, https://commons.wikimedia.org/wiki/File:%22Colored%22_drinking_fountain_from_mid-20th_century_with_african-american_drinking.jpg, public domain.

Without addressing health disparities from the perspective of systemic racism, historical injustices will only be perpetuated.

5.11 Racism and Bioethics

In November 2020, the American Medical Association recognized racism as an urgent and serious threat to public health, after publicly apologizing in 2008 for its own racist practices (Baker 2016; O'Reilly 2020). Scientific interest in racism is growing over the last few years: the number of scientific publications on this topic has sharply increased since 2017 (Hamed et al. 2022). In healthcare ethics discourse however, racism has been a relatively neglected topic until recently (Johnstone and Kanitsaki 2010; Elias and Paradies 2021; Ganguli-Mitra et al. 2022). A search in PubMed (with keyword Racism and Healthcare ethics) produced 492 results. The first publication in 1978 discusses the Tuskegee syphilis study (Brandt 1978), but until 2014 the annual number of publications on the topic was very low (not exceeding ten). Only from 2014 has the total grown (from 14 in 2014 to 106 in 2021). Searching with the keywords Race and Bioethics produced a lower number of publications (often the same as in the other search). The first article in this search addresses the issue of institutional ethics committees and their role in calling attention to problems such as racism (Farley 1984). Here too, publications on this topic are very scarce or altogether lacking for many years, until 2020 when 15 articles appeared, with a maximum of 62 in 2021 and 57 in 2022.

When race realities are discussed in scholarly bioethics journals, they are often not located in a contemporary context but either projected in the past or in colonial Africa (Hoberman 2016). Documentation concerning the inferior care received by racialized minorities is usually not taken into account, and neither is the existing bioethics literature of medical racism. The four editions of the *Encyclopedia of Bioethics* (from 1978 until 2014) present only limited scholarship about racism (Galarneau and Smith 2022). As a reference work that should be a resource for the current knowledge in bioethics, it falls short, not only because articles on racism are rare, but also because it does not incorporate available scholarship, does not contain critical bioethical analyses of racism, and does not highlight racism as a concern of justice (Galarneau and Smith 2022).

Why is racism a relatively neglected topic in bioethics? Johnstone and Kanitsaki (2010) mention four reasons: the failure to examine racism as an ethical issue, the illusion of non-racism in healthcare, the association of awareness-raising of racism in healthcare with whistleblowing, and the sense that the issue is too problematic. The second reason seems the most important one, though various aspects contribute to the idea that racism does not exist in medicine and healthcare.

First, a belief persists that these domains are special and exceptional because they are governed by implicit and explicit normative principles and rules that exclude unfair treatment, disrespecting and harming patients. There has been a reluctance to accept that racist beliefs, attitudes and behaviors are as prevalent among healthcare professionals as in the general public. Second, it is not common knowledge that racist practices cause substantial harm, and even if evidence is presented, it is typically dismissed and discredited (Stone and Dula 2002). Third, even if it is true that health professionals are not explicitly and intentionally racist, the role of systemic racism is hidden and not visible. Structural and organizational racism is often explained with other mechanisms than racial injustice (Elias and Paradies 2021). This is connected to a fourth aspect: racism is explained not as an ethical issue but as a social and political one; it is beyond the scope of bioethics (Galarneau and Smith 2022; James 2022). A fifth aspect is that it is difficult for bioethicists to engage racism as an ethical issue because of the context in which they often operate. Working in a clinical setting, bioethicists are embedded in an existing power structure, making it problematic to criticize discriminatory practices and a healthcare system that sustains inequality. Those working in an academic setting are dependent on funding opportunities related mostly to biotechnologies (Ho 2016). Finally, the prevailing idea that bioethics should be colorblind impedes an ethical analysis of racism since it eliminates the experience of color as a bioethical issue (Galarneau and Snith 2022). The argument is that bioethics is constructed and practiced within an ideological context of "whiteness" (Myser 2003a). Its dominant epistemologies and normative framework, its applications and performances, as well as what is deemed relevant and worthwhile are determined by a particular origin and standpoint marked by White privilege. As long as this positioning within a specific social, and particularly racial hierarchy is not critically examined, issues of race and racism cannot be sufficiently analyzed and resolved.

5.12 The Whiteness of Bioethics

Earlier in this chapter, several aspects of the normativity of colors have been elaborated that might be relevant to the current debate on the color of bioethics. First, it is pointed out that colors usually evoke normative associations. This is true for colors such as red and yellow but especially for white and black. In numerous cultures, black is connected to negative associations, whereas white is the preferred color. These preferences are operative at a general level, regardless of whether colors are attached to specific objects or entities. It might be that implicit racial bias, which is unconscious and unreflective, is first generated by these preferences, whereupon the concomitant associations are then projected onto the person whose skin shows a particular color. This chapter furthermore discussed the history of Western suspicion towards colors, with colors viewed as eliciting emotions, pointing to foreignness, and diverting from rational thinking; as such, colors should not be trusted. The exception was whiteness, which is often not considered as a color; it is colorless and thus opposed to colors (Batchelor 2000). As the absence of color, it is viewed as the norm from which deviations can be assessed (Dyer 2017). When medical doctors and philosophers contributed to the development of racial science, they generally assumed whiteness to be the most original, perfect and beautiful color. Having a white skin is the "natural" and "ordinary" way of being human: it is a neutral, "unracialized" position, transcending embodiment, situatedness and relationality (Dyer 2017).

Against this backdrop, the argument that bioethics is characterized by the "normativity of whiteness" can be understood (Myser 2003a). It does not articulate that bioethics has the same explicit or implicit assumptions as previous racial science; racial beliefs, attitudes and behaviors are categorically rejected as incompatible with any ethical approach. However, at a more fundamental level, bioethical discourse is racialized, i.e. in the moral values and principles that it regards as foundational or universal, and in the subjects of inquiry that are considered relevant and crucial. Almost a decade after Myser's seminal article, Russell reiterates this argument (2022), pointing out that bioethics, having emerged as a new discipline in the 1970s and becoming increasingly mainstream, is based on an underlying principle of White supremacy, i.e. the idea that White lives are of greater value than the lives of people of color. The theoretical framework of bioethics with respect to autonomy, consent, transparency and risk assessment presupposes individual citizens who are independent and free to make decisions, ignoring mostly non-White people who are disadvantaged and vulnerable because of social, economic and environmental conditions. In bioethical analyses, white is usually not considered as a color itself so that White people become invisible as a racial group. The result is that the concept of race is only applied to non-White people (so that the word 'color' becomes equivalent to 'race' and 'non-White'), while Whites are regarded as a social group which is neutral in race relations, and which is also the norm from which deviations are assessed. Assuming that its contents and methods can be determined independently from historical and cultural origins and standpoints, bioethics discourse demonstrates what Dyer (2017) has argued in another context: White is equated with being human, and is the embodiment of universality.

Characterizing mainstream bioethics as 'White' is a serious criticism. It is disconcerting for its practitioners, the majority of whom are White, even if it is argued that the characterization does not refer to the skin color of bioethics professionals but rather to the principles and norms that are promulgated and considered relevant. Whether it is possible to abstract colors from the objects or subjects to which they are attached, depends on the philosophical theory of color one wants to defend. According to color relationism and the phenomenological perspective highlighted in this book, color is a relational experience; it emerges in the interaction between person and world, perceiver and environment. In this perspective it is difficult to imagine Whiteness as an abstract entity, operating apart from White people. Pointing out the dominance of Whiteness in bioethics, Myser (2003a) refers to the fact that its practitioners have been and still are overwhelmingly White. They are the people responsible for the development and construction of the kind of bioethics she labels as "White." They have impregnated bioethics discourse with their (White) values but have made this coloration invisible through presenting it as neutral and cross-cultural. The remedy is that mainstream bioethics is broadened and revised by incorporating the voices and visions of minority populations. It requires that White bioethicists, in particular, recognize and critically scrutinize their own Whiteness (Myser 2003a). These comments show that it is questionable to criticize the discipline whilst letting its practitioners off the hook.

Another complication of this argument about mainstream bioethics is that Whiteness is presented as a static and homogenizing superstructure: White people are treated "as a collective order with a common cultural identity" (Hartigan 1997, 498). The diversity of Whites is not taken into account. This is first of all true for color itself. Colors present themselves in a range with varying hues and intensities. White people are not really white, unless they are very ill or moribund. The same goes for dark colors which present themselves also in a huge diversity. It is directly visible in the Humanae project of Angelica Dass, mentioned in Chapter 1, showing that labels like black, white, yellow or red are inadequate to cover the diversity of the color tone of faces (Dass 2024). It is even more true if the focus is on beliefs and attitudes among Whites; it is hard to identify a common core that is clearly distinct from that of non-White people. Especially from a global perspective, it is doubtful whether there is a shared sense of identity among Whites. If there is White identity, it is furthermore not static but transforming in disparate political and cultural landscapes (Hartigan 1997).

Perhaps the diagnosis of Whiteness first and foremost applies to North-American bioethics, although even that is challenged (Baker 2003). From a global perspective, the situation seems more complicated. While ideas of race and practices of racism persist in numerous other countries, it not only articulates a distinction between black and white but also involves other skin colors or is not primarily related to colors at all. For example, in the Netherlands discrimination is mostly directed against the Moroccan minority population (Veling et al. 2007). In the United Kingdom, Asian communities are the target of racism (Malik 2023). People with albinism are persecuted in some African countries (United Nations 2017). In many places across the world, anti-Asian racism and anti-Semitism have increased, especially during the Covid-19 pandemic (Zack 2023). In a global reference frame, associating bioethics with Whiteness is concurrent with identifying its American-ness. The discipline as it has emerged in the United States displays specific values and concepts which are characteristic of the ethos of this country (Myser

2003b). The problem is that this bioethics interpreted itself as a neutral and universalized discourse, not only appropriate and applicable within the US but also in the wider world. However, it has long been recognized that mainstream North-American bioethics, with its emphasis on analytic philosophy, pragmatism and liberal individualism, is different from other approaches to bioethics, notably European or African bioethics, for example (Ten Have and Gordijn 2001; Tangwa 2019). Labeling bioethics as White can therefore be seen as a call to broaden the scope of bioethics: to focus on cultural, social and economic dimensions of health and healthcare, and on the underlying mechanisms of social injustice and systemic racism (Danis et al. 2016). It is also a call to go beyond the mere analytic approach of clarifying moral issues and quandaries, presenting itself as a "thinking enterprise" whereas it should focus on social change, advocacy and activism (Dula 1991). Furthermore, identifying bioethics as monochromatic is an appeal to pluralism and diversity, including more values in its conceptual and methodological approaches (Truong and Shariff 2021). That means involving multiple disciplines, engaging community (especially minority) perspectives, and voices from various cultures and traditions around the globe. The aim is not to incorporate non-White values in mainstream bioethics so that a new mainstream can be created that is no longer White but rather colorblind (Arekapudi and Wynia 2003). Rather, it involves exploring different value systems, to take a diversity of viewpoints in ethical discourse seriously in order to find shared values.

5.13 Conclusion

In his dialogue *Phaedrus*, Plato imagines the soul as a charioteer driving a team of horses. One of the horses is noble and good, the other has the opposite character—it is crooked, hot-blooded, undisciplined and hard to control. The first horse is white, the second black (Hackforth 1972, 103; 253C-E). For Plato, the soul is the most important part of the human being and it has a tripartite structure. Reason is preeminent, which rules the whole (the charioteer). The other part includes higher emotions (the white horse), and the last part (the black horse) is the locus of desires and carnal lusts.

Plato's allegory illustrates that colors have, besides an aesthetic and emotional role, a normative function; they can be used to express notions of goodness and badness. A specific color may refer to passion and irrational behavior, and should elicit avoidance and control. It may also be used to label treacherous and unreliable people which leads to stigmatization and exclusion. An important use of color is to distinguish and classify entities in the surrounding world. In history there is continuous debate about which colors are suitable in particular circumstances, for specific people and for the expression of social status. This chapter discussed how and why colors are moralized. They are regarded as Plato's black horse, impeding rational thinking. In European culture, at times they are approached with suspicion since they are deceptive. Color is like makeup or an envelope, a second skin that hides what it is covering. It is furthermore marked as foreign and as a sign of otherness.

When the focus is primarily on black and white, it is clear that both are associated with moral views. In a variety of cultures, white evokes purity, hygiene and innocence whereas black is associated with negativity and immorality. These prevailing moral connotations show their impact when people come to be classified on the basis of skin color. The concept of race, introduced in the 17th century, expressed the idea that humans can be categorized according to biological criteria such as skin color. From the beginning, categorization is permeated with normative judgments of the nature, character and temperament of differently colored people, reflecting the moral views that were long attached to various colors, especially in Western culture. However, in racial taxonomies, color proved to be a confusing criterion since it does not clearly demarcate different races. The moral connotations of whiteness and blackness nonetheless prevailed over the actual skin color. When the "white race" was constructed in the United States for example, many European immigrants were initially not included. Differences of color first of all represent moral differences with whiteness as the apogee of rationality, morality and civilization.

The second part of the chapter examines how physicians and philosophers have contributed to the development of racial science and how contemporary societies assume that ideas of race and racism no longer play a significant role in civil life. However, numerous experiences and studies show that racist thinking and practices have not disappeared. Often strongly rejected and morally condemned, racism persist through mostly unconscious and unintentional individual prejudices as well as structural factors that systematically disadvantage non-White people. This is noticeable in the context of healthcare when patients and care professionals report racists attitudes and behaviors. Such reports are frequently denied since the principles of healthcare ethics and the self-image of healthcare providers emphasize impartiality, objectivity and equal treatment. That systemic disadvantages can be embedded in organizational, institutional and structural arrangements of healthcare is difficult to identify and recognize, while action to transform or remediate the resulting injustices is often regarded as a social and political rather than medical and ethical responsibility.

The third and last part of this chapter points out that in the context of bioethics, race and racism are relatively neglected issues. A variety of reasons may explain this lack of attention, but an important one is that in response to the increased sensitivity to racial discrimination, and perhaps to redress its past involvement in racial science, medicine and healthcare have rigorously eliminated the experience of color as a relevant issue. Healthcare providers are outraged when the issue of racial bias and discrimination is brought up, and most bioethicists will not spend much time on it since this is evidently a morally objectionable topic. The effect is that ethical reflection on racism is limited, and that there is no critical analysis of the moral wrongness of the idea of race and racism, and their deleterious consequences for healthcare. The ideology of colorblindness is criticized with the argument that mainstream bioethics in fact is characterized by "whiteness." Its normative frameworks, value systems and epistemologies originated in and are sustained from the perspective of White privilege.

This critique highlights that in current debates about race and racism the focus of attention is shifting from black to white. Whiteness itself has become problematic with criticisms of White superiority and White privilege (Dyer 2017). Forms of human domination that were formulated in the past as White supremacy to justify the ideology of slavery, still endure. It is not recognized that because of a long history of exploitation and injustice, benefits and advantages are accorded to Whiteness, and that health advantages are bestowed to White people because material resources and opportunities are still unequally distributed. The legacy of this history is not recognized since white is regarded as neutral and impartial. Being White itself is not taken as problematic since it assumes that no value judgments are involved. That this is changing is visible on both sides of the political spectrum. Contemporary movements such as "wokeism" are motivated by resistance to the power of White men (Weyns 2023). On the other hand, the great replacement theory, popular among conspiracy and reactionary thinkers, regards White people as an endangered species: this theory argues that White populations are systematically being replaced through mass immigration of non-Whites, and intermingling between White and non-White people (Rose 2022).

White has become a metaphor for a world that is disappearing, while for anti-racists it is a symbol of power, specialness and superiority. The focus on whiteness, however, shares the same prejudices as the pejorative connotations of blackness, attributing negative moral qualities to a specific color. They reflect anxieties and fears about a world changing through globalization, demography, immigration, wars, climate, disparities and structural violence. Both keep alive the ideology of colorism: discriminatory treatment of individuals based on skin color. They forget that it is the power of colors to condition our attitudes, behavior and ways of thinking, at the same time as colors themselves are ambiguous and can induce various associations. Furthermore, they ignore that it is the kaleidoscopic nature of colors that makes the world attractive, enjoyable, beautiful and interesting. Living in a world that is colorless, as the case discussed in the beginning of this chapter illustrates, we would only perceive black and white. Such a world would seem dark and depressing. The question is how bioethical discourse should deal with the issue of color. Should it remain colorblind, reflect a particular color—if not white than black or otherwise?—or appreciate the full range of colors that enhances human existence? That will be the subject of the next chapter.

References

Adams, F. M., and Osgood, C. E. 1973. A cross-cultural study of the affective meaning of color. *Journal of Cross-Cultural Psychology* 4 (2): 135–156.

Allen, T. W. 2021. *The invention of the white race. The origin of racial oppression*. London and New York: Verso.

Arekapudi, S. and Wynia, M. K. 2003. The unbearable whiteness of the mainstream: Should we eliminate, or celebrate, bias in bioethics? *The American Journal of Bioethics* 3 (2): 18–19, https://doi.org/10.1162/152651603766436126

Baker, R. 2003. Balkanizing bioethics. *The American Journal of Bioethics* 3 (2): 13–14, https://doi.org/10.1162/152651603766436090

Baker. R. 2016. Race and bioethics: Bioethical engagement with a four-letter subject. *American Journal of Bioethics* 16 (4): 16–18, https://doi.org/10.1080/15265161.2016.1145302

Balfour-Paul, J. 1998. *Indigo*. London: British Museum Press.

Batchelor, D. 2000. *Chromophobia*. London: Reaktion Books.

BBC. 2023. What do colours on the BBC weather maps mean? *BBC*, July 25, https://www.bbc.com/weather/features/66293839

Bernasconi, R., and Lott, T. (eds). 2000. *The idea of race*. Indianapolis, IN: Hackett Publishing Company.

Blackstock, U. 2024. *Legacy: A black physician reckons with racism in medicine*. New York: Viking.

Blum, L. 2002. *"I'm not a racist, but…" The moral quandary of race*. Ithaca, NY and London: Cornell University Press.

Boswell, D. A., and Williams, J. E. 1975. Correlates of race and color bias among preschool children. *Psychological Reports* 36: 147–154.

Brandt, A.M. 1978. Racism and research: The case of the Tuskegee syphilis study. *Hastings Center Report* 8 (6): 21–29.

Brusatin, M. 1986. *Histoire des couleurs* [*The history of colors*]. Paris: Flammarion.

Calhoun, A. 2021. The pathophysiology of racial disparities. *New England Journal of Medicine* 383 (20): e78.

Challener, C. 2021. How Covid-19 is influencing trends in paint colors. *CoatingsTech* 18 (1), https://www.paint.org/coatingstech-magazine/articles/how-covid-19-is-influencing-trends-in-paint-colors/

Clark-Hitt, R. C., Malat, J., Burgess, D., and Friedemann-Sanchez, G. 2010. Doctors' and nurses' explanations for racial disparities in medical treatment. *Journal of Health Care for the Poor and Underserved* 21: 386–400.

Cullen, D. A., Sword, G. A., Rosenthal, G. G. et al. 2022. Sexual repurposing of juvenile aposematism in locusts. *Proceedings of the National Academy of Sciences of the United States of America* 119 (34): e2200759119, https://doi.org/10.1073/pnas.2200759119

Danis, M., Wilson, Y., and White, A. 2016. Bioethicists can and should contribute to addressing racism. *American Journal of Bioethics* 6 (4): 2–12.

Danso, A., and Danso, Y. 2021. The complexities of race and health. *Future Healthcare Journal* 8 (1): 22–27.

Dass, A. 2024. *Humanae*, https://angelicadass.com/photography/humanae/

DeZIM. 2022. *National Discrimination and Racism report: Rassistische Realitäten. Wie setzt sich Deutschland mit Rassismus auseinerander?* Nationalen Diskriminierungs und Rassismusmonitor (NaDiRa) des Deutschen Zentrums für Integrations- und Migrationsforschung (DeZIM).

Dong, Y., Meng, W., Wang, F. et al. 2023. "Warm in winter and cool in summer": Scalable biochameleon inspired temperature-adaptive coating with easy preparation and construction. *Nano Letters* 23 (19): 9034–9041.

Dula, A. 1991. Toward an African-American perspective on bioethics. *Journal of Health Care for the Poor and Underserved* 2 (2): 259–269, https://doi.org/10.1021/acs.nanolett.3c02733

Dyer, R. 2017. *White*. London and New York: Routledge (20th anniversary edition).

Editorial. 2020. Systemic racism: Science must listen, learn and change. *Nature* 582: 147.

Elias, A., and Paradies, Y. 2021. The costs of institutional racism and its ethical implications for healthcare. *Bioethical Inquiry* 18: 45–58.

Elizabethan Sumptuary Statutes. 2001. Who wears what. Elisabethan.org, www.elizabethan.org/sumptuary/who-wears-what.html

Emery, A. E. H. 1988. John Dalton (1766–1844). *Journal of Medical Genetics* 25: 422–426.

Eze, E. C. (ed.). 1997. *Race, and the Enlightenment*. Malden: Blackwell Publishing.

Farley, M.A. 1984. Institutional ethics committees as social justice advocates. *Health Progress* 65 (9): 32–35, 56.

Feagin, J., and Bennefield, Z. 2014. Systemic racism and U.S. health care. *Social Science & Medicine* 103: 7–14, https://doi.org/10.1016/j.socscimed.2013.09.006

Ford. R. T. 2021. *Dress codes. How the laws of fashion made history*. New York: Simon & Schuster.

Forth, G. 2018. Purity, pollution, and systems of classification. In: Callan, H. (ed.), *The International Encyclopedia of Anthropology*. Chichester: John Wiley & Sons, https://doi.org/10.1002/9781118924396.wbiea2003.

Frank, M. G., and Gilovich, T. 1988. The dark side of self-and social perception: Black uniforms and aggression in professional sports. *Journal of Personality and Social Psychology* 54 (1): 74–85.

Gage, J. 2013. *Colour and culture. Practice and meaning from Antiquity to abstraction.* London: Thames & Hudson (original 1993).

Galarneau, C., and Smith, P. T. 2022. Speaking volumes: The *Encyclopedia of Bioethics* and racism. *Hastings Center Report* 52 (2): S50–S56, https://doi.org/10.1002/hast.1371

Ganguli-Mitra, A., Shahvisi, A., Ballantyne, A., and Ray, K. 2022. Racism in healthcare and bioethics. *Bioethics* 36: 233–234, https://doi.org/10.1111/bioe.13014

Goethe, J. W. von. 1970. *The theory of colours*. Translated by C. L. Eastlake. Cambridge, MA: MIT Press (original 1810).

Gould, S. J. 1996. *The mismeasure of man*. New York and London: W. W. Norton & Company (revised and expanded edition).

Hackforth, R. 1972. *Plato's Phaedrus*. London: Cambridge University Press.

Hall, W. J., Chapman, M. V., Lee, K. M. et al. 2015. Implicit racial/ethnic bias among health care professionals and its influence on health care outcomes: A systematic review. *American Journal of Public Health* 105 (12): e60–e76, https://doi.org/10.2105/ajph.2015.302903

Hamed S., Bradby, H., Ahlberg, B. M., and Thapar-Björkert, S. 2022. Racism in healthcare: A scoping review. *BMC Public Health* 122 (988), https://doi.org/10.1186/s12889-022-13122-y

Hannaford, I. 1996. *Race. The history of an idea in the West*. Washington, DC and London: The Woodrow Wilson Center Press and The Johns Hopkins University Press.

Hartigan, J. 1997. Establishing the fact of whiteness. *American Anthropologist* 99 (3): 495–502.

Harvey, J. 2013. *The story of black*. London: Reaktion Books.

Ho, A. 2016. Racism and bioethics: Are we part of the problem? *American Journal of Bioethics* 16 (4): 23–25, https://doi.org/10.1080/15265161.2016.1145293

Hoberman, J. 2016. Why bioethics has a race problem. *Hastings Center Report* 46 (2): 12–18, https://doi.org/10.1002/hast.542

Hoeppe, G. 2007. *Why the sky is blue. Discovering the color of life*. Princeton, NJ and Oxford: Princeton University Press.

Institute of Medicine (US) Committee on Understanding and Eliminating Racial and Ethnic Disparities in Health Care. 2003. *Unequal Treatment: Confronting Racial and Ethnic Disparities in Health Care*. Edited by B. D. Smedley, A. Y. Stith, and A. R. Nelson. Washington, DC: National Academies Press (US).

Jablonski, N.G. 2012. *Living color. The biological and social meaning of skin color*. Berkeley, CA and London: University of California Press.

Jablonski, N. G. 2021. Skin color and race. *American Journal of Physical Anthropology* 175: 437–447, https://doi.org/10.1002/ajpa.24200

James, J. E. 2022. Race, racism, and bioethics: Are we stuck? *American Journal of Bioethics* 22 (3): 22–24, https://doi.org/10.1080/15265161.2022.2027565

James, M., and Burgos, A. 2023. Race. In: Zalta, E. N., and Nodelman, U. (eds), *The Stanford Encyclopedia of Philosophy* (Summer 2023 Edition), https://plato.stanford.edu/archives/sum2023/entries/race/

Johnstone, M-J., and Kanitsaki, O. 2010. The neglect of racism as an ethical issue in health care. *Journal of Immigrant and Minority Health* 12: 489–495, https://doi.org/10.1007/s10903-008-9210-y

Jones, C. P. 2000. Levels of racism: A theoretical framework and a gardener's tale. *American Journal of Public Health* 90 (8): 1212–1215.

Jones, D. S., Podolsky, S. H., Kerr, M. B., and Hammonds, E. 2023. Slavery and the *Journal*—Reckoning with history and complicity. *New England Journal of Medicine* 389 (23): 2117–2123, https://doi.org/10.1056/nejmp2307309

Kareklas, I., Brunel, F. F., and Coulter, R. A. 2014. Judgment is not color blind: The impact of automatic color preference on product and advertising preferences. *Journal of Consumer Psychology* 24 (1): 87–95, https://doi.org/10.1016/j.jcps.2013.09.005

Kaya, N., and Epps, H. 2004. Relationship between color and emotion: A study of college students. *College Student Journal* 38 (3): 396–405, https://doi.org/10.1353/csd.2004.0062

Lewis, T. T., Cogburn, C. D., and Williams, D. R. 2015. Self-reported experiences of discrimination and health: Scientific advances, ongoing controversies, and emerging issues. *Annual Review of Clinical Psychology* 11: 407–440, https://doi.org/10.1146/annurev-clinpsy-032814-112728

Maina, I. W., Belton, T. D., Ginzberg, S., Singh, A. and Johnson, T. J. 2018. A decade of studying implicit racial/ethnic bias in healthcare providers using the implicit association test. *Social Science & Medicine* 199: 219–229, https://doi.org/10.1016/j.socscimed.2017.05.009

Malik, K. 2023. *Not so black and white. A history of race from white supremacy to identity politics*. London: Hurst & Company.

Marwah, I. S. 2022. White progress: Kant, race and teleology. *Kantian Review* 27 (4): 615–634, https://doi.org/10.1017/s1369415422000334

Maund, B. 1995. *Colours. Their nature and representation*. Cambridge, UK: Cambridge University Press.

Mills, C. W. 2005. Kant's *Untermenschen*. In: Valls, A. (ed). 2005. *Race and racism in modern philosophy*. Ithaca, NY: Cornell University Press, 169–193.

Montagu, M. F. A. 1941. The concept of race in the human species in the light of genetics. *Journal of Heredity* 32 (8): 243–248.

Myser, C. 2003a. Differences from somewhere: The normativity of whiteness in bioethics in the United States. *American Journal of Bioethics* 3 (2): 1–11, https://doi.org/10.1162/152651603766436072

Myser, C. 2003b. A response to commentators on 'Differences from somewhere: The normativity of whiteness in bioethics in the United States'. *American Journal of Bioethics* 3 (3): 56–62, https://doi.org/10.1162/152651603322874825

Nicholson, K. 2022. No, weather maps have not changed colour to make you more afraid of the climate. *HuffPost*, 25 July, https://www.huffingtonpost.co.uk/entry/have-weather-maps-changed-twitter-theory-heatwave-climate-crisis_uk_62d69888e4b0116f21c06717?utm_campaign=share_email&ncid=other_email_o63gt2jcad4

Nobles, M., Womack, C. Wonkam, A., and Wathuti, E. 2022. Science must overcome its racist legacy. *Nature* 606: 225–227, https://doi.org/10.1038/d41586-022-01527-z

Omi, M., and Winant, H. 2000. Racial formation in the United States. In: Valls, A. (ed.), *Race and racism in modern philosophy*. Ithaca, NY: Cornell University Press, 181–212.

O'Reilly, K. B. 2020. AMA: Racism is a threat to public health. *AMA*, 16 November, https://www.ama-assn.org/delivering-care/health-equity/ama-racism-threat-public-health

Painter, N. I. 2010. *The history of white people*. New York and London: W. W. Norton & Company.

Paradies, Y., Truong, M., and Priest, N. 2014. A systematic review of the extent and measurement of healthcare provider racism. *Journal of General Internal Medicine* 29 (4): 364–387, https://doi.org/10.1007/s11606-013-2583-1

Paradies, Y., Ben, J., Denson, N., Elias, A., Priest, N., Pieterse, A., Gupta, A., Kelaher, M., Gee, G. 2015. Racism as a determinant of health: A systematic review and meta-analysis. *PLoS One* 10 (9): e0138511, https://doi.org/10.1371/journal.pone.0138511

Paradies, Y. 2017. Racism and health. In: Quah, S. R., and Cockerham, W. C. (eds), *The International Encyclopedia of Public Health, Vol. 6*. Oxford: Academic Press (2nd edition), 249–259.

Pastoureau, M. 1989. L'Église et la couleur, des origines à la Réforme. [The Church and color, from origins to Reformation] *Bibliothèque de l'école des chartes* 147: 203–230.

Pastoureau, M. 2001. *Blue. The history of a color*. Princeton, NJ and Oxford: Princeton University Press.

Pastoureau, M. 2009. *Black. The history of a color*. Princeton, NJ and Oxford: Princeton University Press.

Pastoureau, M. 2010. *Les couleurs de nos souvenirs* [The colors of our memories]. Paris: Éditions du Seuil.

Pastoureau, M. 2019. *Yellow. The history of color*. Princeton, NJ and Oxford: Princeton University Press.

Pastoureau, M., and Simonnet, D. 2005. *Le petit livre des couleurs.* [The little book of colors] Paris: Éditions du Panama.

Ray, K. 2023. *Black health. The social, political, and cultural determinants of black people's health*. New York: Oxford University Press.

Quillian, L., Heath, A., Pager, D., Midtbøen, A. H., Fleischmann, F., and Hexel, O. 2019. Do some countries discriminate more than others? Evidence from 97 field experiments of racial discrimination in hiring. *Sociological Science* 6: 467–496, https://doi.org/10.15195/v6.a18

Rijksoverheid. 2024. *Uitkomsten Personeelsenquête Rijk (PER) over racisme op de werkvloer en verkenning minder vrijblijvende maatregelen.* [Results of a survey among government personnel on racism on the shop floor and exploration of less optional measures] Kenmerk 2024-0000079362.

Riley, C. A. 1995. *Color codes. Modern theories of color in philosophy, painting and architecture, literature, music, and psychology*. Hanover and London: University Press of New England.

Rose, S. 2022. A deadly ideology: How the 'great replacement theory' went mainstream. *The Guardian*, 8 June, https://www.theguardian.com/world/2022/jun/08/a-deadly-ideology-how-the-great-replacement-theory-went-mainstream

Russell, C. 2022. Meeting the moment: Bioethics in the time of Black Lives Matter. *American Journal of Bioethics* 22 (3): 9–21, https://doi.org/10.1080/15265161.2021.2001093

Sacks, O. and Wasserman, R. 1987. The case of the colorblind painter. *The New York Review of Books*, November 19: 25-34.

Saini, A. 2019. *Superior. The return of race science*. Boston, MA: Beacon Press.

Saito, M. 1996. A comparative study of color preferences in Japan, China and Indonesia, with emphasis on the preference for white. *Perceptual and Motor Skills* 83: 115–128.

Sandford, S. 2022. The taxonomy of 'race' and the anthropology of sex: Conceptual determination and social presumption in Kant. In: Lettow, S., and Pulkkinen, T. (eds), *The Palgrave Handbook of German Idealism and Feminist Philosophy*. Cham: Palgrave Macmillan, 131–150, https://doi.org/10.1007/978-3-031-13123-3_8

Seeman, M. V. 2017. Dress makes the doctor. *Hektoen International* 9 (4), https://hekint.org/2017/01/29/dress-makes-the-doctor/

Sherman, G. D., and Clore, G. L. 2009. The color of sin. White and black are perceptual symbols of moral purity and pollution. *Psychological Science* 20 (12): 1556–1564.

Smedley, A. 2024. Racism. *Encyclopedia Britannica, 29 December,* https://www.britannica.com/topic/racism

Smedley, A., Wade, P., and Takezawa, Y. I. 2024. Race. *Encyclopedia Britannica,* 25 December, https://www.britannica.com/topic/race-human

Stone, J. R., and Dula, A. 2002. Wake-up call. Health care and racism. *Hastings Center Report* 32 (4): 28.

Stuurman, S. 2000. François Bernier and the invention of racial classification. *History Workshop Journal* 50: 1–21.

Stuurman, S. 2017. *The invention of humanity. Equality and cultural difference in world history.* Cambridge, MA: Harvard University Press.

Tangwa, G. D. 2019. *African perspectives on some contemporary bioethics problems.* Newcastle upon Tyne: Cambridge Scholars Publishing.

Ten Have, H. A. M. J., and Gordijn, B. (eds) 2001. *Bioethics in a European perspective.* Dordrecht: Kluwer Academic Publishers.

Truong, M., and Shariff, M. Z. 2021. We're in this together: A reflection on how bioethics and public health can collectively advance scientific efforts towards addressing racism. *Bioethical Inquiry* 18: 113–116, https://doi.org/10.1007/s11673-020-10069-w

Tweedy, D. 2016. *Black man in a white coat: A doctor's reflections on race and medicine.* New York: Picador.

UNESCO. 1952. *The race concept: Results of an inquiry.* Paris: Unesco, https://unesdoc.unesco.org/ark:/48223/pf0000073351

United Nations. 1965. *International convention on the elimination of all forms of racial discrimination,* https://www.ohchr.org/en/instruments-mechanisms/instruments/international-convention-elimination-all-forms-racial

United Nations. 2017. *Report of the Independent Expert on the enjoyment of human rights by persons with albinism on the Regional Action Plan on Albinism in Africa (2017–2021).* Human Rights Council, Thirty-seventh session, 26 February–23 March 2018; A/HRC/37/57/Add.3, https://digitallibrary.un.org/record/3846088?v=pdf

United Nations. 2021. OHCHR and racism. *OHCHR,* February, https://www.ohchr.org/en/racism

Valls, A. (ed.). 2005. *Race and racism in modern philosophy.* Ithaca, NY: Cornell University Press.

Van Dijk, T. A. 1999. Discourse and racism. *Discourse & Society* 10 (2): 147–148.

Veling, W., Selten, J. P., Susser, E., Laan, W., Mackenbach, J. P., and Hoek, H. W. 2007. Discrimination and the incidence of psychotic disorders among ethnic minorities in The Netherlands. *International Journal of Epidemiology* 36 (4): 761–768.

Weyns, W. 2023. *Wie Wat Woke. Een cultuurkritische benadering van wat we (on) rechtvaardig vinden*. [Who What Woke. A cultural critique of what we find (un)just] Kalmthout: Pelckmans Uitgevers.

WHO Commission on Social Determinants of Health. 2008. *Closing the gap in a generation. Health equity through action on the social determinants of health*. Geneva: WHO.

Williams, D. R., and Mohammed, S. A. 2009. Discrimination and racial disparities in health: evidence and needed research. *Journal of behavioral medicine* 32 (1): 20–47, https://doi.org/10.1007/s10865-008-9185-0

Williams, D. R., and Wyattt, R. 2015. Racial bias in health care and health. Challenges and opportunities. *Journal of the American Medical Association* 314 (6): 555–556, https://doi.org/10.1001/jama.2015.9260

Wilson, J. 2024. Scientist cited in push to oust Harvard's Claudine Gay has links to eugenicists. *The Guardian*, 14 January, https://www.theguardian.com/world/2024/jan/14/christopher-rufo-jonatan-pallesen-eugenics-racism-claudine-gay-harvard

Woo, T., Liang, X., Evans, D. A. et al. 2023. The dynamics of pattern matching in camouflaging cuttlefish. *Nature* 619: 122–128, https://doi.org/10.1038/s41586-023-06259-2

Yin, R., and Ye, H. 2014. The black and white metaphor representation of moral concepts and its influence on moral cognition. *Acta Psychologica Sinica* 46 (9): 1331–1346, https://doi.org/10.3724/sp.j.1041.2014.01331

Zack, N. 2023. *Ethics and race. Past and present intersections and controversies*. Lanham: Rowman & Littlefield.

6. A Colorful Bioethics

6.1 Introduction

During the Covid-19 pandemic, attention was called to the fact that the disease is associated with a particular color. Right from the start of the pandemic, it was evident that the disease has a disproportionate impact on African Americans: they have higher rates of infection, hospitalization and deaths from Covid-19 than the White population of the United States. This disparity encouraged a search for biological factors that could explain racial differences, such as blood type or gene expression. The assumption that special African American vulnerability to Covid-19 must be due to innate biological differences demonstrates the continuing power of the idea of race in medicine and healthcare (Xue and White 2021). It exemplifies a historic line of thinking that Black bodies are different, concluding that specific 'Black diseases' can be identified. However, the color of Covid-19 is not black. Other racial and ethnic minorities are also disproportionately involved. In Australia and New Zealand, for example, Indigenous populations are more affected than non-Indigenous people (Elias and Ben 2023). Furthermore, it is evident that in the United States and elsewhere, the incidence and severity of the pandemic disease is associated with higher levels of risk among certain populations, rather than their color itself. Since minorities already experience structural health inequalities, have more co-morbidities, and less access to care, the virus has more impact on their health. During the pandemic, there was also an upsurge in racism and racist hate crimes across a wide range of countries, manifested as discrimination and violence against Asian people and foreigners (Elias and Ben 2023).

These experiences of the recent pandemic are important lessons when we ask the question of how bioethics should deal with color. The first is that the idea of race itself is a bioethical issue. While the idea has no scientific reality, its continuing and often implicit use in medical theories and practices should be exposed and critically examined, and bioethical discourse should scrutinize how the idea operates in various contexts (Russell 2022). The second is that racism can no longer be a neglected topic in bioethics discourse. Bioethics should deconstruct the power differences in racist discourses, practices and structures (Johnstone and Kanitsaki 2010). The third lesson is that bioethics needs a wider ethical framework that will enable it to address and eliminate the deleterious influences of color on health and healthcare.

6.2 Race as Bioethical Issue

The concept of race continues to be used in contemporary clinical medicine, health research, medical guidelines and standards of care. The assumption is that categorizing patients and research participants into racial or ethnic categories is relevant for adequate diagnosis, treatment and care. Human diversity cannot be denied. Since humans are embodied beings, it is important to identify biological and genetic differences between human populations. The conclusion therefore is that "race can be a medically useful category" (Lorusso and Bacchini 2023, 452). This conclusion, however, is uncomfortable and problematic since the concept of race itself is morally problematic. Thinking in terms of race has, according to Blum (2002), four consequences. First, it is inherently divisive. Instead of what all human beings share and have in common, it focusses on fundamental and permanent differences between people, creating moral distance among human populations, and promoting ideas of "otherness." Second, it assumes that people classified in the same race share common characteristics, so that less attention is paid to the diversity within racial groups. This not only leads to stereotyping but also hinders engagement with the patient as a unique individual within their specific setting, since attributes of the group are assigned to each individual member of the group. Because the idea of race implies that there are different kinds of people with different essential characteristics, the third consequence is that these characteristics are immutable and inescapable. Finally, races are identified and classified not as neutral descriptions based on physical appearance such as skin color, but rather, every racial classification is evaluative. The color terms used to refer to races mobilize the existing normative associations within Western culture. They reflect prevailing value judgments concerning inferiority and superiority, for example, the belief that the Caucasian face is the best approximation of perfect beauty.

Blum's distinction of the moral dangers of racial thinking explains how race conflicts with the ethical framework of bioethics. It is not simply that the idea of race is harmful and has deleterious results for numerous people. The idea also defies important bioethical principles; it not only contradicts the moral principles of equality and justice but also the principle of respect for personal autonomy and responsibility. It is incompatible with the notion of human dignity which requests that all human beings deserve respect and recognition. Blum's distinction furthermore clarifies what makes the idea of race attractive, and thus persistent, especially in a medical context: it focuses on biological and genetic constitution rather than socio-economic conditions. If racial categories are features of the natural world, science can discover and describe those properties and make them relevant for medical treatment. Interpreting itself as an objective science, medicine primarily examines and elucidates the physiological basis of conditions that affect various racial groups. It assumes that this scientific approach does not imply a value judgment though concerns exist about possible misuses of this objective information. It is taken for granted that the use of the notion of race in research, medicine and biotechnology is not harmful in itself, and that potential abuses can and should be prevented. This assumption, however, is false since what motivates scientists to search for scientific explanations of racial differences is the fundamental belief that races exist. However, there is no scientific evidence that they do exist, and genomic studies show that genetic variation within racial populations is greater than between racial populations. For example, when it is argued that obesity in Black women has a physiological basis due to a genetically determined metabolism, this is not founded on an objective, value neutral hypothesis but presupposes the idea of race, i.e. the belief that Black bodies are essentially different, and have "innate" genetic abnormalities compared to White bodies (Tsai et al. 2020).

The continuing use of racial categories furthermore illustrates that racial categorizing is attractive since it favors a particular type of explanation. While obesity is a complex phenomenon, the focus is exclusively placed on biological, and specifically genetic explanations, disregarding the extensive literature on the influences of the social environment. Evidence that healthy food and food security are severely restricted because of racial inequities in income, wealth and distribution of community resources is not taken into account. Neither are the physiological effects of chronic stress due to interpersonal and systemic racism. This example illustrates that the persistence of the idea of race is linked to a particular self-interpretation of medicine as an objective natural science focused on biological explanations. Moreover, what the example clarifies is that racial distinctions have a particular practical purpose. While in the past, the category of race was used to classify conquered peoples, and to justify the control, domination, exploitation or enslavement of others, it is still useful nowadays to serve specific purposes. According to Roberts, applying this category to human beings introduces a political division: "… a system of governing people that classifies them into a social hierarchy based on invented biological demarcations" (Roberts 2011, x).

Classifying people according to race has consequences. The implication is that differences between humans cannot be ameliorated or eliminated through social policies and programs. Health and disease depend on physiological differences, on innate strengths and weaknesses. Health disparities therefore do not require social and economic policies but rather biological interventions: "A chief reason why genetic explanations are emphasized over social ones is that genetic causes can be treated with a pharmaceutical product" (Roberts 2011, 146). Racial classifications, even when they are no longer based on externally visible characteristics such as skin color, but are located at the molecular level—beneath the skin, as biological or genetic differences—are still associated with ideas of privilege and deprivation, superiority and inferiority, normality and abnormality, and thus have differential practical implications (Zack 2023). This is evidenced, as will be discussed below, in the idea that people of different races suffer from specific diseases. When Black bodies are viewed as biologically inferior to White bodies (which are taken as the norm, and control group), specific Black diseases can be distinguished (Yearby 2021). This is evidenced, for instance, in the idea that Black bodies have a higher tolerance of pain, resulting in reduced need for anaesthesia in surgery, or at least lower dosages of pain medication (Akinlade 2020; Ray 2023). It can also be seen in contemporary diagnostic algorithms and practice guidelines, with the practice of adjusting outcomes according to race (Vyas, Eisenstein and Jones 2020).

Given these consequences, it has been advocated that the concept of race should be eliminated in medicine and healthcare: the only way to end racism in healthcare is to stop using all references to race (Blum 2002; Yearby 2021). Particularly in some areas, such as medical genetics, the use of concepts of race is problematic and harmful and should be discontinued (Yudell et al. 2016). The counter-argument posits that the notion is medically useful, not because race as a biological category exists, but because it is indispensable to study the effects of racism on health, for example in epidemiological studies of the consequences of racism for health and health inequalities. Racism is a fundamental cause of health inequalities, either directly or indirectly through its impact on social-economic status, and it cannot be addressed without the notion of race. Thus, paradoxically, the concept is necessary to fight systemic racism in healthcare (Lorusso and Bacchini 2023). However, this strategy is precarious since the continuing use of the notion can strengthen the belief that biological and genetic differences are racially determined. If the notion is used it should be well-considered, explained and critically examined, avoiding any suggestion that there are hierarchical differences among human groups.

Nonetheless, the problem remains that a standard definition of race in medical, epidemiological or health services research is lacking, and that racial groups are heterogenous and not clearly demarcated (Braun et al. 2007). In biomedical publications it is often not clear how and why information on race (and also ethnicity) is collected (Kaplan and Bennett 2003). While many geneticists assume that human biology differs by race, they have difficulties in defining race and are unclear about the meanings of the race categories used (Fullwiley 2007). Reports on associations between genetic phenomena, health outcomes and race (or ethnicity) often do not mention how the race (or ethnicity) of research participants is determined (Shanawani et al. 2006). The most common method to assign race is self-identification, but it is not always clear whether it is on the basis of offered and preselected response choices or self-reporting (Moscou et al. 2003). Not all people identify with a single racial category, and self-identification may evolve over time so that responses vary in different contexts and will be difficult to compare (Kaplan and Bennett 2003). Lastly, collection of data on race and ethnicity in healthcare settings may be perceived as harmful by minority and marginalized patients themselves (Varcoe, et al. 2009). In response to these challenges, some have proposed the study of human diversity and commonality using other concepts than race (such as ancestry and geography) (Yudell 2016). But it is doubtful whether this will prevent the interpretation of these differences in racialized terms (Bradby 2012).

6.3 Race under the Bioethical Microscope

The conclusion so far is that the continued use of race in medicine should be regarded as a morally problematic practice. Bioethical discourse must therefore not just identify where and how the concept is employed but critically explore the reasons for its use, and seriously scrutinize it. This will require a shift in bioethical thinking and a critical stance towards routine practices of clinical medicine and biomedical research (James and Iacopetti 2021). Such a shift will contribute to the transformation of race-based medicine into a race-conscious, and ideally race-free medicine.

First of all, critical bioethical discourse should focus on the language used in clinical and research settings. Already in 1998, Bhopal and Donaldson argued that racial labeling of patients is misleading, inaccurate and superficial. Such labeling is often done on the basis of color with "white" referring to people who do not belong to specific racial and ethnic groups and who represent normality in comparison to others. They contended that we should abandon the use of terms like "White," "Caucasian," "Western" and "Asian" (Bhopal and Donaldson 1998). More than two decades later, racial classification and race names continue to be used in medical publications (Jablonski 2021). The label "Caucasian" has been used more than 7,000 times per year in publications since 2012; it was used 6,814 times in 2018, and 6,991 times in 2023 (according to a PubMed search in May 2024). Medical and science journals have recently updated guidance for the use of language and terminology referring to race and ethnicity (Flanagan et al. 2021). The term Caucasian should not be used as synonymous with White, but only to refer to people from the Caucasus region.

Fig. 6.1 Caucasian Biosphere Reserve in the vicinities of Sochi, Russian Federation. Photo by SKas (2016), Wikimedia, https://commons.wikimedia.org/wiki/File:Caucasian_Biosphere_Reserve.jpg#/media/File:Caucasian_Biosphere_Reserve.jpg, CC BY-SA 4.0.

Many other language indications are provided, for example racial and ethnic terms should be used in adjectival rather than noun form (e.g. Asian women rather than Asians). Giving more attention to language may not only reduce unintentional bias in scholarly literature but also reflect the significance of fairness and equity. The effect of editorial language policies remains to be seen. An example of change in medical terminology is Down's syndrome. Working in the 1860s with children with learning disabilities, John Langdon Down noticed that they shared a common appearance. On this basis he classified this congenital condition as "Mongolian idiocy" or "mongolism." Referring to Blumenbach's racial taxonomy, he assumed that these children were a regression or degeneration towards a lower race (Gould 1996, 164–165). At the end of the 1950s when the genetic cause of the syndrome was discovered, geneticists and physicians started to argue that another name should be used (e.g. trisomy 21) in order to avoid racial connotations. The use of the term "mongolism" diminished progressively (Rodríguez-Hernández and Montoya 2011). Since the term "mongol" is derogatory in many languages, the government of Mongolia asked the World Health Organisation to revise the naming, whereupon the Organisation in 1965 decided to abandon the term (Howard-Jones 1979). In 1975, Medical Subject Headings replaced the term with 'Down's syndrome'. However, the term "mongolism" continues to be used, especially in publications since 1980 (rising from 306 publications in 1980 to 2,054 in 2022, according to a PubMed search in May 2024). This example shows that racial terminology persists and that bioethical scrutiny of language will remain necessary.

Another area where the idea of race is influential concerns the perception of diseases and medical conditions. There is a long history of associating race with disease conditions and of regarding some diseases as specific to certain racial groups. A classic example is sickle-cell anemia, which was regarded as a "Black disease" in the United States, where it is most common among African Americans. This was viewed as evidence of essential differences between Black and White groups. Later discoveries showed that the sickle-cell gene provides some resistance to malaria; the difference is geographical and due to natural selection. The same gene is found in populations with different skin shades, and sickle-cell disease exists in Arab countries and India. In some parts of Africa where malaria is not prevalent, the incidence of sickle-cell anemia is lower. While in the US, there is a statistical correlation between Blackness and this disease, the explanation is not race (Blum 2002). Similar correlations have been found between asthma and hypertension: in these cases too, the idea has been put forward that racial groups have structural vulnerabilities due to their biological or genetic constitution, rather than accepting that these groups have higher exposure to harmful socio-economic conditions and systemic racism (Saini 2019). The high mortality of tuberculosis among Black populations, especially after the Emancipation Proclamation in the United States, was attributed not to deleterious living and working conditions, but rather to a racial predisposition, with the lung capacity of Black people presumed to be reduced and thus inferior (Braun 2014).

A more recent example is schizophrenia, which was constructed as a "Black disease" in the 1960s (Metzl 2009).

Another idea is that diseases manifest differently in Black bodies. Indeed, the rationale behind the Tuskegee Syphilis Study was the presumption that syphilis was a different disease in Black people: the view was that this group was especially prone to venereal diseases and also indifferent to treatment, thus anti-syphilitic treatment was unnecessary (Brandt 1978). The assumption that certain diseases are peculiar to specific races continues to influence contemporary medicine. Cystic fibrosis is underdiagnosed in African Americans today because the general idea is that this is a typical disease of White people (Rubin 2021). The same is true for autism. The stereotypical image of this disorder is that it primarily affects White children. In the United States, Black autistic children receive an accurate diagnosis years later than other children, although they present the same clinical symptoms (Fombonne and Zuckerman 2022). Delays in diagnosis lead to later access to quality intervention services (Constantino et al. 2020). In the Netherlands, the same is true for people with a diverse immigration background (Morocco, Turkey and Surinam). If they manifest atypical behavior, it is often attributed to their cultural origin (Stift 2024).

The idea of race is furthermore observable in the clinical setting. The notion that racial groups have different lung capacity and function, and that in particular Black people have lower lung function than White people, produced the practice of "race correction" or "ethnic adjustment." Since the value for normal lung capacity is based on measurements for White bodies, the values for other racial groups are adjusted and mostly reduced. Braun, who studied the history of the spirometer, the main instrument to measure lung function, shows how race correction is already programmed in the instrument itself (Braun 2014). Since "non-White" lungs are assumed to be inferior to the standard of normality, the spirometer can "objectively" demonstrate why they are predisposed to respiratory disease. It also means that disability claims in case of occupational hazards, for example in mining, are limited. The emphasis on innate differences in lung capacity implies that social and environmental factors can be ignored. The employment of separate standards in clinical medicine, based on the idea of racial differences, has come under increasing scrutiny, and has become a significant issue on the agenda of bioethics. It is evident that the establishment of "normal" values since the early nineteenth century has been based on studies of White populations (Crenner 2014), with Black subjects rarely included in research populations. The resulting White normality is now increasingly recognized as biased. In June 2023, for example, the American Medical Association decided that the body mass index (BMI) is an imperfect measure of healthy bodyweight. It is established on an imagined ideal of the "average" man, based on a sample of White, European men. Used for racist exclusion, it cannot predict the risk of disease on an individual level across different racial and ethnic groups (Berg 2023). Race norming is apparently embedded in numerous diagnostic algorithms and practice guidelines, predicting, for example, that Black patients are less likely to have a kidney stone, or have lower risk of osteoporosis. Such race-adjusted algorithms have harmful consequences since further evaluation of complaints may be postponed, and diagnosis and intervention delayed (Vyas et al. 2020). Measures used to calculate kidney function, including an often-criticized algorithm, are adjusted because they suggest that Black people have better kidney function. This implication often results in delayed referrals for specialist care or transplantation (Braun et al. 2021). Using different standards of normality in clinical medicine to account for disparities in health outcomes in racial populations may be justified if they are the result of genetic differences. But this is, according to Vyas and colleagues "exceedingly unlikely" (Vyas et al 2020, 879). In fact, adjustments presuppose the idea of races that are inherently dissimilar. Racial disparities in health outcomes most likely reflect the effects of racism. "Race correction" will only perpetuate and exacerbate the existing disparities since it will impede access to clinical services (Neal and Morse 2021). Artificial intelligence is now increasingly used in medicine, but since it relies on historical data based on biased data generation or clinical practices, the risk is that existing biases are propagated (Parikh et al. 2019).

Biomedical research is another area where the notion of race continues to be used. In order to improve the inclusion of persons from racial and ethnic minority groups in clinical trials, many guidelines have been issued since the 1990s. In the US, regulatory bodies and funding agencies such as the National Institutes of Health request the use of racial and ethnic categories in clinical research (generally using the categories of the Office of Management and Budget of the US Census Bureau) with the aim of expanding diversity and inclusion in research (Callier 2019). These classifications are now widely used in other countries, although terms and approaches to define population subgroups may vary considerably. In the United Kingdom, for example, eighteen options are offered with five main groups, including "Arab" and "Indian." While in the United States, race is distinguished from ethnicity, this is not the case in other countries. Most countries (65%) enumerate their populations by national or ethnic group, and only 15% employ the notion of race (mostly New World societies with histories of slavery). The term "nationality" is mostly used in countries in Eastern Europe and the former Soviet Union (Morning 2008). In countries such as France, Germany and Spain, ethnic classification as well as the use of "race" is legally prohibited (Gombault et al. 2023).

This diversity of classifications is problematic, especially for global research and cross-country comparisons. Terminologies and criteria differ substantially, and labels for racial and ethnic groups are often simplistic, but this is not merely a methodological problem, as Morning (2008) argues. Populations are mainly classified because of political concerns, and the debate about the (in)appropriateness of such classifications is primarily a normative one. Classification schemes are divisive, stigmatizing and marginalizing. They also promote the interpretation of human diversity in terms of biological and genetic differences, distracting from the examination of other determinants of medical conditions (Gombault et al. 2023). The basic ethical quandary is reflected in the ongoing debate about the use of racial and ethnic categories in biomedical research. On the one hand it is argued that such categories should be abandoned since they perpetuate the idea of race as an explanation of human diversity (Fullilove 1998; Yudell et al. 2016). On the other hand, the argument is that without reporting these categories, the underrepresentation of minority populations in research cannot be addressed, health inequities and different risk profiles cannot be understood, and effective health policies cannot be designed (Burchard et al. 2003; Dessie and Chen 2023). Although this last argument is ethically motivated, and certainly not racist, the risk is that it will sustain the idea of race as an explanation of human diversity. That this risk is not imaginary is shown in a conceptual analysis of recent publications about Covid-19 in which there is a prevailing tendency to biologize the categories of race and ethnicity (Malinowska and Žuradzki 2023). Despite the recognition that these categories correspond to genetic, socio-cultural and environmental differences, they are most often solely explained in terms of biology and genetics. Social and environmental factors influencing differences in health are usually ignored and not explored. The categories of race and ethnicity are not simply described but used as explanatory tools to analyze human diversity. Biogenetic reductionism may also clarify the fundamental paradox that the limitations and inadequacies of racial and ethnic designations are commonly known in the research community but nonetheless extensively used and reported. Although researchers desire to eliminate racial health disparities, biomedical research continues to promote a biogenetic rather than social interpretation of racial variation (Gutin 2019).

The last area where the idea of race persists is medical education. Examination of the content of courses in the preclinical medical curriculum shows that race is often misrepresented as a biological category. The language used to discuss health disparities may be imprecise and antiquated; racial and ethnic differences in disease burden are presented without context and critical discussion of underlying causes, such that they are attributed exclusively to genetic predisposition; race is portrayed as risk for disease, linking diseases to racial groups, and pathology in general to race; race-based clinical guidelines are taught without questioning their interpretation and evidence (Amutah et al. 2021). The various ways in which race is used in medical education therefore reinforces already existing implicit biases among students and physicians, and also makes the systemic racism embedded in biomedical approaches of health and disease invisible. Students have argued that medical education should be reformed in order to have a more critical evaluation of race (Tsai et al. 2016; Nieblas-Bedolla et al. 2020). What is needed is an examination of the historical and social context of race-based medicine (Braun and Saunders 2017). Programs to eliminate racial bias are necessary but not sufficient; attention should also be given to systemic racism, and to the tools and measures connecting this to negative health outcomes (Futterman et al. 2024).

This overview of where and how the notion of race is used in medical language, in labeling disease conditions, in clinical guidelines, in biomedical research and in education demonstrates how critical bioethical thinking can contribute to transforming race-based medicine. First, it is not sufficient to argue that the term "race" should not be used; it has generated and structured much of the scientific knowledge with which medicine operates. It is an ethical responsibility to critically reassess the scientific basis for many contemporary theories and practices. Given the history of abuse, and the many imperfections in the use of racial categories, the evidence of past research involving racial variables should be re-examined and validated, while for future investigations the first question should be why such variables are used at all (Ioannidis et al. 2021). Second, employing the notion of race implies categorization and homogenization; it diverts attention from the unique situation of individual patients. Codes of medical ethics generally declare that the primary duty of physicians is to promote the health and well-being of the individual patient. The consequence is that it is important to know first of all the patient's history, family history and social context instead of assessing his or her race (Braun et al. 2007). To diagnose patients and understand differences in disease risk, it is better to work from symptoms, history and context than from racial assumptions (Futterman et al. 2024). Third, the notion of race persists due to the domination of biomedical perspectives that give priority to biological and genetic explanations. Too little attention is often paid to structural, social and cultural mechanisms that shape medical knowledge, so that the root causes of illness, and the sociopolitical and historical foundations of health inequities are not considered. This is often the result of lack of interdisciplinary approaches, as is particularly manifested in scientific research and medical education (Braun and Saunders 2017). The remedy is not only training in cultural competency and in awareness of bias and prejudice—since these are primarily focused on changing individual attitudes and behaviors—but emphasis should also be placed on social determinants of health. This requires "structural competency": the ability to discern how mechanisms and forces in society produce health inequities, social vulnerabilities and poor patient care (Metzl and Hansen 2014). The implication is that bioethical analysis should not only use a microscopic but also a macroscopic perspective.

6.4 Racism as Bioethical Issue

As explained in the previous chapter, the concept of race emerged in the seventeenth century, whereas the term "racism" has only been in use since the early twentieth century.. This recent history, however, does not imply that the phenomenon of racism did not exist before the term was coined. The practice of racism can be traced back to the late medieval and early modern periods of European history (Fredrickson 2002). Following the atrocities of the Second World War, racism was generally condemned and prohibited in national and international legislation. However, it is remarkable that in most cases and documents the term is not defined. For example, the Council of Europe set up a European Commission against Racism and Intolerance in 1993 to monitor action against racism, discrimination, and intolerance in Europe, but it does not describe what is regarded as racism (Council of Europe 2024). In response to questions from the European Parliament, the European Commission defined racism as "Ideas or theories of superiority of one race or group of persons of one colour or ethnic origin" (European Commission 2024). In the United Kingdom, the Equality Act 2010 relates racism to less favorable treatment on the basis of race, skin color and ethnic or national origin (Gov.UK 2015). According to the Dutch government, racism is a theory, idea or opinion implying a subdivision of human beings on the basis of presumed race, and considering one or more groups as superior or inferior (Ministerie van Binnenlandse Zaken en Koninkrijksrelaties 2022). The most extensive description is provided in the *UNESCO Declaration on Race and Racial Prejudice*: "Racism includes racist ideologies, prejudiced attitudes, discriminatory behavior, structural arrangements and institutionalized practices resulting in racial inequality as well as the fallacious notion that discriminatory relations between groups are morally and scientifically justifiable" (UNESCO 1978). As argued in the previous chapter, this formulation highlights two characteristics of racism: inferiorization and antipathy. It clarifies that racism concerns not only an ideology or worldview but practices and behaviors resulting in differential treatment, stigmatization, marginalization, exclusion and discrimination.

Race and racism are closely connected. It is evident that the first is presupposed in the second. Racism assumes the belief that people are intrinsically different because they have innate and unchangeable characteristics due to a specific biogenetic constitution or ethnic identity. The assumption that removing any references to race would eliminate racism is too simplistic. For decades, it has been argued that races have no biological or genetic reality; but racism persists as long as people assume that they exist, even if it is clear that they do not exist. Moreover, the idea of race is not merely an individual belief but is embedded in practices, structures, organizations and policies, as the previous section of this chapter has demonstrated. Most people in contemporary societies do not endorse the idea of race, and will reject explicit and implicit racial bias. Nonetheless, systemic racism subtly, covertly and unconsciously sustains the significance of race. Scholars have therefore argued that race is, in fact, the product of racism (Roberts 2011; Bonilla-Silva 2022). As long as the racial ordering of the world continues, race will remain a relevant notion. This is the reason why colorblind policies—pretending not to see the color of somebody's skin—are inadequate to eliminate racism. While it is crucial to deconstruct the notion of race and its uses in medical settings, it is equally, perhaps more, important to morally denounce, challenge and eliminate all forms of racism at all levels in healthcare.

Racism as an ideology and practice influences relations among human beings as well as the functioning of human societies. It promotes certain ideas about racial purity, superiority and inferiority with practical consequences for how human beings live together. Ethical discourse is concerned with the quality of human co-existence, and reflects on what ought to be done to make the human condition better, or at least to counteract its deterioration. Against this backdrop, racism is definitely an ethical issue since it negatively impacts how human beings live together. There are several arguments why racism is morally wrong from the perspective of bioethics.

The most common argument is that racism is harmful. Through direct and indirect discrimination, implicit biases in personal interactions, and systemic racism embedded in structures, organizations and practices it harms people physically and psychologically—and within the context of healthcare, patients in particular. Such harms are preventable if attitudes, beliefs and behaviors, as well as structures and systems were not racist. However, harmful effects of racism also have a specific characteristic since they are not incidental but occur in a more permanent and durable manner. Harms arise "in enduring ways" since they are the result of historical legacies which have created disparities, socio-economic and power differences that continue to determine how people relate to each other even today (Johnstone and Kanitsaki 2010, 491). For bioethics it is relevant that racism produces harm at three levels (Russell 2022). Health is dependent on a range of physical, social and environmental conditions such as employment, healthy food, safe living conditions, educational opportunities and unpolluted environments. However, some groups of people experience negative influences of socio-economic determinants of health because of residential segregation, lack of decent employment, food deserts, less access to education and public transportation, and contaminated drinking water. Reduced capacities and resources for health are associated with socio-economic status, but even when socio-economic status (and individual behavior) is taken into account, disparities between White and Black populations continue to exist in the United States (Yearby 2021).

Reducing health disparities should also attend to racism as a source of inequality, particularly the negative health effects of pervasive discrimination. The harm of racism is also manifested at the level of healthcare. Earlier, many examples were given of lack of access to healthcare and lower quality of care for racial minorities. Such disparities are often explained in terms of health illiteracy, lack of cultural competency and socio-economic context, but not in reference to racism. Finally, racism is harmful at the level of the healthcare system. Russell argues that it explains the resistance to the development of more equitable and inclusive healthcare systems in the US. Neoliberal policies and the ideology of free-market competition perpetuate stereotypes that portray some groups, particularly White individuals, as motivated and responsible, while depicting others, especially non-White individuals, as lazy, dependent and irresponsible (Russell 2022). This argument illustrates that the harms of racism not only impact racialized groups but everybody in need of equal and competent healthcare (Yearby 2021).

Another bioethical argument that racism is objectionable emphasizes the ethical principle of justice. Racism implies differential and unfair treatment, and is therefore unjust because it violates the notion that all humans should be equally treated. On this basis, racial discrimination is nowadays explicitly prohibited. When racial attitudes, beliefs and behaviors still occur, they are usually considered as incidental and exceptional, and as symptoms of prejudice and implicit bias. However, this interpretation of justice as equality does not address systemic racism. Racialized minorities face unjust social arrangements as a result of historical systems of oppression, domination and exploitation which have produced deep inequalities in wealth, political power, employment and educational opportunities which still exist today. In order to create a just society, it is not sufficient to treat people equally (on the basis of the notion of equality) but it is necessary to provide equal opportunities (on the basis of the notion of equity). Rather than equal treatment, it is necessary to remove inappropriate and unjust barriers that obstruct people of color more than others in society (Shelby 2014). It is important therefore to recognize different concepts of justice to counter racism. Frequently used is the concept of distributive justice: equal distribution of harms and benefits. It acknowledges that different parties are in unequal positions and aims to establish a fair distribution of goods and services. But the underlying causes of maldistribution of resources are usually not taken into consideration. Resources must be distributed fairly, but why has the need to (re)distribute arisen in the first place? That injustice is more than maldistribution is, for example, argued in environmental justice movements (Ten Have 2019). Dumping toxic waste in minority neighborhoods is wrong because it disregards the health and well-being of racialized minorities and does not respect them as citizens. Without addressing the context of oppression and inequalities of power, the interests of racialized groups are ignored, and fair distribution of resources will not ameliorate the injustices. Respect and dignity are preconditions for distributive justice.

Given these critical considerations, the concept of social justice is better applicable in relation to racism. It highlights the social structures and mechanisms that produce systemic racism. Instead of discussing issues of access to healthcare and distribution of resources for those who are harmed, it primarily accentuates how people are made vulnerable and how their health is negatively impacted. Focusing on the social and institutional conditions that produce inequalities will enable critical analysis of the power differences and inequal structures that make racialized groups more vulnerable than others, and will also provide insights into how these structures can be transformed and remedied. The perspective of social justice furthermore directs attention to the fact that the racial structure of society disadvantages racialized minorities and benefits the White majority. Existing social structures, practices and relations reinforce White privilege, which is regarded as unjust since it results from a historical legacy of slavery, colonialism and exploitation (Bonilla-Silva 2022). A third concept to remediate racial injustices is restorative justice. This concept articulates that it is more important to identify who is hurt by racism and who is accountable and obligated to amend the harm done, than to search for who or what is to blame and deserves retribution. Addressing and repairing the harms of racism requires recognizing the suffering of individuals and understanding the ongoing, damaging influences of the past (Minow 2022). Such healing efforts can target both interpersonal and systemic racism. Only recently (July 2023), the King of the Netherlands officially apologized for the country's role in slavery. In several countries, slavery monuments and museums have been established, memorializing the history and legacy of slavery and the slave trade. Commemorative events, such as Juneteenth in the United States, mark the end of slavery, while the United Nations General Assembly designated March 24 as the annual International Day of Remembrance of the Victims of Slavery and the Transatlantic Slave Trade.

Fig. 6.2 Alex da Silva, *Slavery Monument* (2013), Rotterdam. Photo by GraphyArchy (2020), Wikimedia, https://commons.wikimedia.org/wiki/File:GraphyArchy_-_Wikipedia_00706.jpg#/media/File:GraphyArchy_-_Wikipedia_00706.jpg, CC BY-SA 4.0.

A third bioethical argument against racism is rooted in the principle of human dignity. The notion of race originated during the Enlightenment as a means to justify persistent inequalities in Western European societies, where equality was proclaimed as a moral and political ideal (Malik 2023). The discrepancy between ideal and practice was bridged by categorizing people into racial groups and associating different qualities with each race. This idea of race was attractive since it not only explained inequality but was also actionable: if some groups are deemed subhuman, they are not entitled to the same treatment as those considered fully human. Moreover, if differences are based on race, they are permanent and inalterable, and policies to change the social context or educational efforts are futile. The basic moral problem with this view is dehumanization: members of a subordinated group do not have the same moral standing as others (Shelby 2014; Bonilla-Silva 2022). This clearly violates the notion of human dignity, which posits that all human beings have intrinsic dignity and equal moral worth. While this concept is not exclusive to Western thought, with deep roots in various religious and cultural traditions (Andorno and Pele 2016), the modern notion of human dignity emerged in the mid-twentieth century. Before that time, more limited conceptions were advanced, such as the idea of moral dignity in Western Antiquity (emphasizing that humans are capable to develop moral ideas and virtues), and spiritual dignity in Christian theology (the human person as created in the image of God) During the Enlightenment, the idea of dignity was related to rationality.

Nonetheless, these ideas of dignity were limited and not universally applied to all human beings. For example, as discussed in the previous chapter, Immanuel Kant accepted a fundamental difference between White and Black races, assuming only the White race to be capable of moral progress, such that moral agency is primarily a characteristic of White European men. This changed in the twentieth century with the growing significance of the human rights discourse, resulting in the *Universal Declaration of Human Rights* (1948). Article 1 of this Declaration states that "All human beings are born free and equal in dignity and rights" (United Nations 1948). This transformed the concept of human dignity not only into a universal moral principle but also a legal and political one, incorporated into national and international legal documents. It has also been accepted as an overarching principle in modern bioethics.

For the ethical context, it is important to notice that human dignity has two aspects. As an abstract and theoretical notion, it applies equally to each individual human being, regardless of characteristics or conditions such as race, age, gender or (dis)ability. It is an intrinsic quality that does not depend on whether it is respected or recognized, or whether it is disregarded by authorities or political systems. The other aspect is that human dignity is a practical phenomenon, a lived experience (Bieri 2017). It is not merely an intrinsic but also a relational quality, since it refers to how humans are and should be treated. Human beings can experience how their dignity is disrespected, denied or lost. Social science research clarifies that the most salient experiences of racism are those of being disrespected, underestimated and ignored (Lamont 2023, 65). The principle of human dignity emphasizes that persons and things are different. Because they have intrinsic and equal dignity, human persons are subjects who should be respected and protected; they should not be used as things that can be owned, exploited for various purposes, or exchanged as commodities. The principle protects subjects against objectification, inferiorization and exploitation. Racism is therefore morally unacceptable since it negates the two aspects of human dignity: denial of the intrinsic moral worth of all human beings, and alteration of human interaction into experiences of humiliation, disrespect, lack of recognition and discrimination. Both aspects derive from the awareness that all human beings share fundamental needs and vulnerabilities, making human dignity the basis for mutual respect and care in decent societies across the world.

6.5 The Color of Bioethics

In dealing with issues of race and racism, bioethical discourse generally follows the policy of colorblindness that has become prevalent in many Western countries. The assumption is that over the past sixty years, due to the civil rights movement, social welfare policies and stringent legislation, racial discrimination and inequality have ended: the idea of race has become obsolete and racism discredited. Since the law requires that individuals should be treated similarly, racial identity is irrelevant and race should not play any role in social life. Against this background, it is preferable not to see color, demonstrating that racism is history. When instances of racism nevertheless occur, they are explained from the perspective of individualism. Persistent racial discrimination is due to the biases and prejudiced attitudes of some individuals, and continuing racial inequalities are the result of failure and lack of effort of individuals who do not take responsibility for their lives. The general change in normative climate in the 1960s and 1970s that made overt and explicit racism unacceptable and instigated the rise of colorblind policies has also resulted in the emergence of bioethics. In this new discipline, respect for personal autonomy has become one of the most important ethical principles. It articulates the rights and responsibilities of individuals as well as the value of individual choice. Race and racism do not play a significant role in bioethics discourse because ethical principles such as respect for personal autonomy but also justice (and equal treatment) provide a normative perspective that makes these notions irrelevant. Colorblindness as the prevailing policy in Western society is thus reflected in bioethics, and may be the reason for neglect of race and racism in bioethical discourses and practices.

Since the turn of the millennium, colorblindness has faced growing criticism. Numerous studies show that racism is not merely a relic of the past but continues to play a significant role in many societies, although in a more subtle and covert way (Neville et al. 2013; Bonilla-Silva 2022; Brown et al. 2023). There is a serious difference between what people say about its unacceptability and their actual behaviors, as explained in the previous chapter. More importantly, critics argue that colorblindness is based on a specific understanding of racism. It regards racism as an individualized phenomenon, and associates it with prejudices and behaviors of individual persons, but does not consider it as a systemic phenomenon, produced by "systems of advantages and exclusion that generate privilege for one racially defined group at the expense of another" (Brown et al. 2023, 43). Most importantly, pretending not to see colors results in denying racial inequalities and discrimination in societies, and in ignoring the different practical experiences of racialized groups. Critical discussions of skin color, racism and discrimination are avoided because the dominating belief is that everybody is similar and equal. By asserting that everyone is equal, colorblindness effectively overlooks racism and therefore sustains inequalities.

Another view is that bioethics should be race-conscious and accept that it has a particular color, namely white. This is the argument discussed in the last chapter. Bioethics should recognize that it has emerged in a particular social and cultural context that takes the norms and values of White people as a self-evident starting point. It should interrogate the underlying assumption of White privilege in its normative framework. Rather than being blind to colors, bioethics should name and recognize differences between people, starting with an acknowledgment of its historically situated perspective. Here, Whiteness refers not merely to the skin color of practitioners but to a cultural norm that influences its normative framework. This shift requires reversing the usual focus: instead of primarily emphasizing the deprivation and discrimination of people of color, attention must also be directed to how the White population is systematically advantaged.

The idea that bioethics has or should adopt a particular color is gaining popularity. In April 2024, the University of Bristol organized the first conference on Black and Brown in bioethics (University of Bristol 2024). There are also advocates of green bioethics, focusing on environmental values and the impact of healthcare practices on the environment (Richie 2018). What these differently colored notions of bioethics have in common is that they direct attention to issues that are relatively underdeveloped in the current theory and practice of bioethics. They generally do not argue in favor of a more narrow or particular view of bioethics focused on one specific issue or theme, resulting in what has been called "balkanization" of bioethics (Baker 2003). What the attribution of specific colors to bioethics illustrates is the necessity of a broader perspective that does more justice to the diversity of people and viewpoints in bioethical theory and practice. Labeling bioethics as White can be taken up as a call to incorporate in bioethical discourse and practice voices, values and visions of populations other than White. It can furthermore be regarded as an appeal to expand moral criticism beyond the perspective of the individual person, and to focus on the cultural, social, political and economic context of health and healthcare, and the underlying systemic mechanisms that produce injustice, inequity and vulnerability—just as labeling bioethics as green refers not to skin color but to the need to attend to environmental concerns that are relevant for health and healthcare. The question posed by these differently colored notions of bioethics is whether they can sufficiently address the socio-cultural and ecological challenges of diversity and difference as they exist across the world. This question has motivated the recent emergence of global bioethics. The confrontation with a range of ethical issues in various contexts as well as divergent normative standards in different parts of the world has instigated a search for shared values and common ideals, while at the same time recognizing that not every global citizen is the same and neither has equal power to cope with the challenges of health and healthcare. In this global perspective, bioethics should be conscious of racism and its consequences, and thus colorful, rather than colorblind or associated with a specific color.

6.7 Ethics and Aesthetics

The theory of color relationism, presented in Chapter 2, postulates that colors are situated between the objective and subjective world. They are first experienced before they are analyzed and interpreted. Quintessentially, they directly address our emotions and feelings because they evoke immediate associations and meanings. Perceiving colors is not merely an observation of the surrounding world but at the same time a normative and aesthetic experience. Taking seriously this typical character of colors has implications for the conception and methodology of ethics.

Bioethics is commonly regarded as a conceptual and abstract system of moral principles and rules, elaborated in ethical theories and codified in legal statements and guidelines. Its general method is to analyze moral dilemmas on the basis of rational reflection and arguments, using clear and transparent procedures for decision-making in medical practice. Although facts and values in healthcare are often emotionally charged—for example, with fear, sadness, anxiety, grief and also disgust, indignation and moral outrage—emotions and feelings are usually regarded as obstacles to rational analysis; they must be controlled before the proper deliberative process can take place. This conception of bioethics is nowadays increasingly criticized (Ten Have 2016; Ten Have and Pegoraro 2022). One argument is that principles require interpretation and cannot directly be applied to moral problems in order to provide clear-cut answers. For instance, when a treatment is recommended, an assessment should be made of possible benefits and harms. But what is beneficial or harmful might be different from the perspective of the patient or the healthcare provider. If the patient does not want a treatment that is clearly beneficial, the healthcare provider should balance the ethical principles of respect for patient autonomy, beneficence and non-maleficence. Another argument is that ethical decision-making is not an abstract and decontextualized process but always takes place within a specific and concrete context and clinical practice. Because human beings are necessarily situated, ethical reasoning draws heavily on the moral experiences of the persons involved. A more fundamental criticism is that moral judgments and decisions are not merely rational but influenced by values and emotions which determine what is morally relevant and significant. Before a moral judgment can be made and before moral reasoning and rational deliberation can take place, situations with which we are confronted must be perceived as important from a moral point of view. These critical views point to the crucial role of moral perception in ethical discourse. Perception requires moral sensitivity. It is also facilitated by the moral imagination that enlarges perspectives and that situates the moral subject in the specific conditions and concrete circumstances of other people. Before engaging in rational analysis, it is important to articulate why a particular situation, experience or problem ethically matters, why it affects us as moral beings.

The role of moral perception in ethical discourse highlights the connection between ethics and aesthetics as the science of sensory perception (Macneil 2017). The ancient proverb *verum, pulchrum et bonum convertuntur* ("truth, beauty and goodness are interchangeable") expressed that, in traditional philosophy, truth, beauty and goodness refer to the same underlying reality, and that science, aesthetics and ethics are thus interconnected. In modernity, these domains of human activity are usually separated. Ethics is practical reasoning concerned with what is good and right. Its aim is to determine what ought to be done with the help of moral principles to guide rational arguments and deliberations. Aesthetics is concerned with beauty and it involves the senses rather than rationality, especially when colors are concerned. Because senses are considered to be less reliable than reason, aesthetics is regarded as a matter of affection and intuition, thus personal taste. Science aims at approximating the truth of reality with empirical methods and with systematic reasoning based on facts and data. However, the distinction is problematic and inconsistent. For example, in science the separation from ethics is contested since scientific theory and practice is influenced and shaped by values and normative presuppositions (Ratti and Russo 2024). At the same time, scientific theories, methods and results are frequently chosen, preferred and presented on the basis of aesthetic criteria such as simplicity and elegance (Derkse 1992). The connection between science and aesthetics is also evident in eighteenth-century racial taxonomies. For Blumenbach, the White race comes first because it is the most beautiful, and for Kant this race has the most beautiful body. Conceptions of beauty at that time were based on classical art, with white marble Greek and Roman statues as the primary examples. It was also the reason why Blumenbach used the term "Caucasian", a region he associated with the most beautiful people. Aesthetic preferences continue to play a role in present-day "colorism" as a basis for racial discrimination. Cultural preferences for lighter skin are motivated not only by ideals of beauty but also social advantages associated with whiter skin tones. These ideals have generated a widespread global practice of skin bleaching (Hunter 2007; Jablonski 2012). In many countries, cosmetic preparations are advertised and used with the promise that lighter skin may bring relief from discrimination, and contribute to social advancement, as if a dark skin is a disease to be cured.

Fig. 6.3 Skin-whitening product in supermarket in Sri Lanka. Photo by Adam Jones (2014), Wikimedia, https://commons.wikimedia.org/wiki/File:Fair_and_Handsome_-_Skin-Whitening_Product_in_Supermarket_-_Bandarawela_-_Hill_Country_-_Sri_Lanka_(14122094934).jpg#/media/File:Fair_and_Handsome_-_Skin-Whitening_Product_in_Supermarket_-_Bandarawela_-_Hill_Country_-_Sri_Lanka_(14122094934).jpg, CC BY-SA 2.0.

That ethics and aesthetics can be connected is furthermore argued by Bueno Pimenta and Garcia Gomez (2023). In their view, the organization of global ethical principles in the *UNESCO Universal Declaration on Bioethics and Human Rights* is a display of beauty. Contemplating these principles is like an aesthetic experience since it transcends cognitive relationships and reveals the various dimensions of being human and the possibilities of human improvement. Aesthetics is not simply picturing, detecting or seeing but evaluating, seeking to see the world anew. It moves from seeing to seeing differently. It is "a general engagement with value, and so it is an ethical undertaking" (Noë 2023).

The term aesthetics is derived from the Greek *aisthánomai* which means perceiving, feeling and sensing. What is beautiful or ugly has an immediate sensory presence, unlike what is true or good. Aesthetic impressions and judgments are based on human sensitivity, imagination and intuitions, and as such assumed to be possible sources of error. Colors provide a standard example: they can be attractive or repulsive, warm or cold, and immediately evoke certain feelings and emotions. But they can also be deceitful; depending on the context they may be illusory or concealing. Colors are like a skin that covers an underlying reality. In the distinction usually made between ethics and aesthetics a similar contrast between the profound and the superficial seems to be at work. Proceeding from rational arguments and deliberation, ethics is the search for goodness, and it identifies reasons for and against acts and decisions that should ultimately convince everyone and provide justifications independent from personal beliefs and preferences. Aesthetics, on the other hand, aims to understand the nature and appreciation of beauty. The aesthetic experience, for example when we view a painting or listen to music, is first of all subjective, eliciting emotions and feelings, or influencing our mood or attitudes. The pleasure that we feel is immediate, and not the product of conceptual thought, analysis or reasoning. Detecting and evaluating are entangled in the aesthetic experience (Noë 2023).

Conceiving ethics as a rational and deliberative activity has currently become problematic. Cognitive psychology research shows that there are in fact two cognitive processes for making a moral judgment. One is a reasoning process as exercised in moral deliberation. It is a conscious process that is analytical, controlled and unfolds in subsequent steps. This is the process generally presented as the paradigmatic method of bioethics. However, there is a second process that, in practice, is more frequently used: it is an intuitive process which is immediate and automatic, based on feelings and emotions, and operating more quickly than reasoning and deliberation. Empirical studies demonstrate that most moral judgments are made through this intuitive process (Haidt 2001). It is therefore problematic to argue that moral judgments are the product of reasoning; they are, in most cases, more correlated to moral emotion than to moral reasoning. At the same time, moral reasoning is used but after a moral judgment has been made. While emotions trigger an intuitive response and result in an immediate moral judgment, moral analysis and deliberation start *post hoc* to offer justifications to others for our moral judgments. This point of view offers a new appreciation of emotions and feelings in bioethical discourse, and as such obliterates the usual separation of ethics and aesthetics. Emotions and feelings should no longer be regarded as obstacles to rational and disinterested moral decision-making but are a necessary and prior ingredient of moral judgments. Moral perception and sensitivity are at least comparable to aesthetic experience; they not only determine what is morally significant but they also provide, so to speak, the "material" for subsequent rational analysis and deliberation.

When we admire a painting and tell a friend that it is beautiful, she might ask why. When we are uneasy, and perhaps angry, with the conduct of a health professional who does not show respect, we comment to a friend that such behavior is wrong and should not happen. Again, she may ask us why. In both examples, an intuitive judgment is directly made which then triggers an exchange leading to a reasoning process. Aesthetic and moral experience are comparable in the sense that both accentuate the role of emotions and feelings in judgments about beauty and goodness. Simultaneously, it is clear that the analytic reasoning that clarifies the judgments made differs. In the first case, the friend might answer that beauty is subjective, a matter of taste, and that she does not like the painting at all. In the second case, a similar answer would not be satisfactory. If our friend points out that the health professional is a nice and competent person who has no intention of offending us, we will feel not be taken seriously. Our emotional response is motivated by moral concerns; something has happened that ought not to happen, whether or not the person involved is nice or competent. The reasons we provide for our uneasiness and indignation go beyond the level of psychological interaction or individual taste; they refer to what is wrong and unacceptable in any interaction of this kind in the setting of healthcare, whoever is involved and wherever it takes place. The intuitive judgment that this is not how healthcare providers and patients should interact with each other is justified with reasons and arguments that apply to human interaction in general. At this *post hoc* level moral deliberation is used to analyze the various aspects of the problem, to distinguish facts and values, and to identify the relevant ethical principles.

It is important to note that moral reasoning and deliberation take place in a social context. We develop arguments in response to the comments of our friend, and in general when confronted with the reasons of other persons. Facing the perspective of others, we might actually change our intuitive judgments. Ethics necessarily is a social activity, which is not evident for aesthetics. Perceptions of beauty may differ among persons, and it may be difficult to convince another person that they should appreciate a painting because I find it so beautiful. In the search for goodness, views may also differ, but at least arguments can be exchanged as to why certain behaviors are good, desirable, commendable or not. These arguments do not express my personal preferences but appeal to what is good for all human beings. Ethical discourse assumes that human beings are not isolated, self-reliant individuals but social beings, connected with others, embedded in social and cultural contexts, and sharing common interests. With moral reasoning and deliberation, humans try to identify the values that they share and that provide a common framework for society. The capacity for moral deliberation is "a kind of social cement" that binds groups together because it confronts the first-person perspective with the perspectives of others and appeals to common perspectives (Christakis 2019, 409).

If moral experience is like aesthetic experience in the sense that it immediately and automatically implies emotional responses producing an intuitive moral judgment, the significance of color in bioethical discourse should be re-evaluated. The prevailing assumption that color must not play any role in normative assessment is the conclusion of moral reasoning. But this conclusion follows after a moral judgment has already been made on the basis of an immediate intuitive response to color. We have discussed in earlier chapters how colors in general evoke immediate associations which are then analyzed and reflected upon. Colors are associated with specific virtues such as honesty, rationality or dignity, or on the contrary, with different vices. This is particularly true for black and white. Colors express meaning and significance; they symbolize normative values. This has been visible in dress codes used to indicate social status and class. They were also used in a more negative way to stigmatize and exclude others from normal social life. Colors therefore play also a social role since they have a particular purpose of ordering, identifying and classifying human and social environment. In this sense they help us to orientate ourselves in the world. The symbolic value and functional role of colors are clearly noticeable when colors are connected to the idea of race. Their meanings are projected onto human beings through their skin, influencing perceptions of their appearance; and colors have been functionally used to construct taxonomies of human races based on judgments of superiority and inferiority. Studies show that the implicit evaluative associations with the colors white and black are systematically correlated with evaluative racial associations (Smith-McLallen et al, 2006). The suggestion is that color preferences form the primary basis for racial preferences, and may well precede racial biases since they are learned early in life.

The main conclusion earlier from historical analysis of the influence of colors is that meanings and associations are changeable. They are not so much determined by the colors themselves, as well as mediated and reinforced in social and cultural interactions and learning processes. The significance of colors has changed over time, as well as in different cultures. During the Reformation, black was regarded as the most dignified and sincere color, while in the Victorian era white was considered as the epitome of beauty. The color blue was, for a long time, not appreciated in Western cultures. Viewed in Ancient Rome as the color of barbarians, in the Middle Ages it became a divine color. Yellow, on the other hand, was adored in European Antiquity when it was associated with gold, the sun, and energy, power and joy. It was also omnipresent in daily life in Eastern cultures. In China, the home of the Yellow Emperor, it was reserved for the emperor. In the Middle Ages in the West, it was transformed into a symbol of treason, deception, envy, jealousy and dishonesty and used to stigmatize and exclude people such as heretics, prostitutes and the mentally ill from society. It also referred to Jewishness and the synagogue. Medical practice amplified the bad reputation of yellow: doctors used uroscopy to diagnose diseases, and associated yellow with liver disease, malaria, pus and putrid fluids, and mold. The chromophobia of Protestant reformers made yellow almost disappear from public life. In Islamic cultures yellow was similarly unfavorable, being considered the color of lying and treachery, as well as of disease and ageing. In surveys on color preferences in the West, it has been the least popular of the basic colors consistently since the 1880s (Pastoureau 2019).

Changes in the normative appreciation are furthermore evident for the color green. In Western cultures, green was for a long time disliked and regarded as a bad color, associated with the devil, monsters and ghosts. Its association with poison grew with the use of green pigments used in paint, which often contained arsenic. In Islamic countries, green was always a positive color, referring to paradise. For Protestant reformers, green was to be avoided as a frivolous and immoral color; it is the color of avarice. Only in the Romantic era, and the second half of the eighteenth century in Europe, did green become a more dignified color, as a symbol of life, vitality and renewal, focusing attention on nature, and as a sanitary color, pointing to health and hygiene. For scientists and painters in the nineteenth century, green became the opposite of red. Nowadays, green has strong moral connotations: ecological responsibility, sustainable development, concern for biodiversity and nature, and protection of the environment (Pastoureau 2014). Similar changes in normative associations have taken place historically and culturally in connection to the colors white and black, as discussed previously.

The normative meanings and values of colors are changeable because they are the result of social and cultural processes. They are learned since early experiences in social and cultural settings that associate specific colors with particular moral qualities. They are also expressed in language and communication. Since value systems of cultures change and because various cultures interact and exchange values with each other, the meanings of colors have been transformed.

The upshot is that when the meanings of colors are the result of social learning processes, their normative associations can be unlearned and transmuted. This is particularly relevant in regard to racial biases. When evaluative associations with white and black are learned and reinforced, and subsequently connected to racial preferences, they can be influenced by cultural, educational and linguistic practices. For bioethical discourse, this means that we should reach behind the denial or trivializing of racial biases due to the general assumption that such biases should not play any role at the explicit level of moral reasoning and deliberation. Bioethical analysis should focus on the emotional level of automatic, unintentional and unconscious processes at which color associations and implied normative evaluations arise, and that produce immediate moral judgments. Implicit associations and biases should be brought to the surface and made explicit and conscious, and we should analyze how they operate, even if we are not aware of their influence. "Automatic thinking" is now an important subject of cognitive sciences, showing how unconscious associations and implicit racial and ethnic bias are malleable (Blair et al. 2001; Rudman et al. 2001; Burgess et al. 2007; Matthew 2015). Mesman, for example, argues how associations between black and bad, white and good, can be toned down when people are made aware of the ways in which our language use evokes and reinforces certain associations. Alternative stories and images also contribute to changing the automatic associations of colors since they enrich the image of the "other" (Mesman 2021). What is needed are narratives and images that empower people, recognizing the value of their experiences, and articulating inclusion, "extending dignity to all groups" (Lamont 2023, 113).

6.8 Moral Imagination

When it is concluded that ethical reflection and moral deliberation are not entirely rationalistic processes but connected to intuitions and emotions, bioethical analysis should focus on the intuitive stage in which moral judgments immediately and automatically emerge. Methodologically, this will require not only moral reasoning but moral sensitivity and moral experience in order to understand why and how we perceive particular situations as morally significant and relevant. But it also requires moral imagination. This is the ability to detach ourselves from our actual situation, taking us beyond the limitations of our empirical experiences. The French philosopher Gaston Bachelard (2014) celebrates the imagination as a creative faculty which allows human beings to surpass and escape reality as given.

6. A Colorful Bioethics 191

Fig. 6.4 Gaston Bachelard (1965), Dutch National Archives, The Hague. Photographer unknown, uploaded by Anefo, Wikimedia, https://commons.wikimedia.org/wiki/File:Gaston_Bachelard_(kop)_filosoof,_Bestanddeelnr_917-9599.jpg#/media/File:Gaston_Bachelard_(kop)_filosoof,_Bestanddeelnr_917-9599.jpg, CC0.

Imagination empowers us to empathize with others because it enlarges our horizon and widens our sympathies, helping us to recognize situations that demand moral reflection and action because we become aware of values that go beyond the limits of our own experience as well as the moral demands that others place upon us.

The essential purpose of imagination in bioethics can be clarified with what traditionally has been called the moral point of view. Ethics as a human activity exists because our sympathies are limited. At the same time, not everyone has similar sympathies. The object of moral evaluation is to contribute to the amelioration of the human predicament. Moral discourse seeks "to countervail 'limited sympathies' and their potentially most damaging effects", in other words, to mitigate the ill effects of indifference of persons to other persons (Warnock 1971, 26, 149). In order to reduce the potential of conflicts, we are encouraged to take the point of view of other persons. In the history of ethical discourse, the circle of moral concern has expanded: more and more beings are taken into account as morally relevant. Against this background, the crucial feature of ethics is the ability to shift perspectives. Even if we start from our own intuitions and emotions to develop a first-person point of view, our ethical life is shared with others obliging us to take into consideration second-person perspectives, initially from people to whom we are attached. The dynamics of social interaction stimulates reflexivity because we recognize that the point of view of others is different from our own. Ethical sensibilities and intuitions are formed through intersubjectivity and reciprocity: we share and exchange perspectives with one another and we are sensitive to the perspective of others. For example, the notion of dignity is not just an individual quality but emerges from interactions; there must be other persons who respect my dignity. Viewing ourselves through the eyes of others initiates a process of taking a distance on ourselves, expanding our moral sensibilities, and produces ethical reflection, generating a third-person perspective with explicit and generalized norms, reasons and justifications that apply outside our immediate sphere of interaction. According to this cognitive development model, the third-person perspective does not provide a complete understanding of ethical life since it excludes first/second-person perspectives, but neither do the last two perspectives (Keane, 2016). What is crucial is the capacity to move back and forth between perspectives. This capacity is provided by moral imagination. In distinction to the imagination that is crucial in aesthetic concerns, moral imagination has a particular direction; it is focused on the perspectives and interests of other persons than ourselves.

In moral analysis and deliberation, moral imagination is essential for two reasons. First, it is necessary for the required shift in perspectives because it enlarges our horizon, expands our sympathies, and helps to frame situations differently (Ten Have and Pegoraro 2022). Through imagination we can place ourselves in the shoes of other people in very different circumstances; we become aware of values that go beyond the limits of our own experience, and recognize situations that demand moral action. Without imagination, we cannot consider situations from the perspective of other persons and cannot understand the experiences of others.

The second reason is that imagination has creative power in that it provides various possibilities for acting, envisioning how actions might be damaging or beneficial, and how alternative courses of action are possible. Ethical reasoning and deliberation are discursive practices that move from specific cases and situations to more elaborate and abstract arguments. In this movement, imaginary processes play a role so that richer and broader views may emerge. Imagination is important to facilitate interpretation, and to generate values, ideals and worldviews to guide moral perception and action. In this way, imagination reshapes and restructures our moral experiences. It helps to understand the situations and views of other human beings, but also to envision how these might be altered and ameliorated. As the ability to make the absent become present, it conceives of alternatives to problematic situations and views. In the philosophy of John Dewey, imagination implies seeing the actual in the light of the possible (Fesmire 2003).

In bioethics theory, practice and education, moral imagination does not seem to play a role. It tends to be regarded as subjective and non-rational, and should therefore be avoided. Publications that emphasize the significance of moral deliberation as a method of ethics teaching in medical education do not refer to the imagination (Steinkamp and Gordijn 2003; Molewijk et al. 2008; Barilan and Brusa 2013). This situation is different in other areas of ethics teaching, for example, education for nursing students, engineering students and student teachers (Jantzen et al. 2023; Jalali, Matheis and Lohani 2022; Hyry-Beihammer et al. 2022). Recent studies in these areas examine the key contribution of the imagination to moral reasoning and deliberation. Jantzen and colleagues (2023) describe how a pedagogical space for the development of moral imagination can be created through simulated learning experiences. Nursing students were trained as simulated patients to confront problems of workplace violence and moral distress. Acting as an angry family member allowed them to imagine the perspective of the patients and their families as well as to identify possibilities to prevent violent situations. They also could critically reflect on the responses of the other students with whom they interacted. This experience transformed their understanding and stimulated them to imagine alternative ways of engaging with patients (Jantzen et al. 2023).

Cultivation of the moral imagination is an important component of moral analysis and deliberation. Imagination is not a subjective and irrational process that can strictly be separated from moral reasoning. Moral deliberation can be viewed as "expansive, imaginative inquiry into possibilities for enhancing the quality of our communally shared experience" (Johnson 1993, 80). Imaginative exploration and transformation of experience can be systematically encouraged in teaching programs through the use of literature, art, movies, role plays, hypothetical and sometimes bizarre cases, and active learning processes (Ten Have 2018; Gerrits et al. 2023). Imagination is, moreover, an effective tool to moderate stereotypes. In experiments conducted by Blair and colleagues, participants who engaged in counter-stereotypic mental imagery (imagining a strong woman, for example a business executive or athlete) produced substantially weaker stereotypes concerning women compared with participants who did not engage in mental imagery (Blair, Ma and Lenton 2001). Implicit bias and prejudice apparently can be reduced with the help of the imagination of counter-stereotypes. Rather than advocating policies of colorblindness to avoid or suppress stereotyping, activation of the imagination is an effective means to reduce and moderate implicit associations with color. Making people aware of implicit race bias and using the imagination as a strategy to reduce bias (e.g. by taking the perspective of stigmatized others, and imagining counter-stereotypic examples) could produce long-term reductions in implicit race bias (Devine et al. 2012).

6.9 Expansion of Bioethical Discourse

The experience that the world is full of colors influences our relations with other people and our environment. Colors present the surrounding world in specific ways and pervade our interaction and communication with other beings. They convey particular emotions, values and judgments, and therefore influence our beliefs, attitudes and behaviors. So far, it has been argued that the typical character of colors has implications for the conception and methodology of bioethics. Perceiving a color or range of colors produces an immediate and intuitive response which generates a value judgment prior to moral reasoning and rational deliberation. Ethics already starts in the concrete experience of perceiving which then necessitates critical examination and explanation with the help of systematic theory and moral reflection. Bioethical analysis should therefore begin with scrutinizing how associations and intuitions emerge, and explore the role of moral perception and moral imagination, especially in regard to ideas of race and practices of racism.

However, there is also the upstream level of bioethical theory where principles, rules and norms are formulated and elaborated that are consolidated in guidelines and legal documents. Recognizing that colors are associated with moral appreciations has consequences for this theoretical framework and its concomitant practices. It not only requires that topics such as racism, structural violence, vulnerability and discrimination should be higher on the agenda of contemporary bioethics, but it demands that the field of ethical inquiry should be expanded by employing a broader framework of ethical approaches and principles.

In Western moral philosophy, human beings are usually conceived as rational and abstract actors, divorced from bodies, feelings and emotions. Rational choice theory assumes that human beings are chiefly concerned with self-interest, motivated by minimizing costs and maximizing gains for themselves. The rational individual makes choices according to what they prefer or value most. They should achieve self-management, i.e. showing responsible conduct and self-regulation. This individualistic ideology, dominating economic and social policies, is reflected in the common view of bioethics. Through their bodies, humans are situated in the world as independent selves, acting on their surroundings. This individual autonomy should be respected. The life of an individual belongs to themselves. The individual person chooses their values, and has the right to live as they would like, being their own master. The moral vocabulary of bioethics is therefore limited: focused on individual rights, self-determination, consent and privacy, rather than social responsibility, solidarity, cooperation and social justice. According to this approach, bioethics should be colorblind, since the prevailing moral principles apply to every individual regardless of race, color and gender. Yet, erasing color as a relevant ethical consideration removes the possibility of exploring why disadvantages and injustices exists, and of analyzing why and how people are treated differently. Ignoring color and pretending "not to see" it does not eliminate difference in reality, and especially differences among people, such as disparities in health and healthcare. Colorblindness, as Anderson argues, is "epistemologically disabling: it makes us blind to the existence of race-based injustice" (Anderson 2010, 5).

As a normative standard for law, policy or ethics, colorblindness accommodates and reinforces the dominating individualism in Western societies. If racist acts and opinions occur, they are regarded as anomalous and unacceptable, and the involved persons will be blamed. But systemic, institutional racism that is embedded in organizations, structures and policies will not be addressed (Neville et al. 2013). Anderson (2010), for example, argues that segregation is the principal cause of racial inequality, providing numerous examples in the area of housing, employment, education and healthcare. Racial prejudices and biases are the effect, rather than the cause of segregation, and reducing or moderating them will not eliminate inequality, stigmatization and discrimination without eradicating the underlying structures of segregation. The consequence of this critique of colorblindness is that ethical analysis should be orientated towards contextual and structural conditions rather than focus on the individual perspective of rational and autonomous persons. Bioethics discourse should transcend the usual emphasis on the moral principle of respect for autonomy. Since racism is the expression of power differences, critical attention should be directed at the power constellations and structures that determine the social, economic, political and environmental conditions in which people live (Johnstone and Kanitsa 2010). Racism is also the production and exploitation of differentiated vulnerability since power disparities deprive racialized groups from the resources required to ensure and maintain health (Russell 2022). Like other ideologies, racism is a dehumanizing system of oppression, domination and exploitation that legitimizes unjust social relations (Shelby 2014). The notion of power is therefore a crucial concept in an expanded bioethical discourse aimed at addressing issues of race, racism and color. The second concept is diversity. Evidently, human beings are not identical; they show significant differences with tremendous biological and cultural diversity. The challenge is to recognize and respect differences without leading to inequality. Racial theories explain differences by categorizing humans on the basis of biology or genetics with the result that differential treatment is necessary and that inequalities are regarded as natural and ineradicable. The paradox is that the ideal of equality which is core to modern societies since the Enlightenment only applies to a specific section of the population. The history of racism and colonialism is, in fact, a corruption of the Enlightenment legacy, as Frantz Fanon (2021) argues. But this legacy should not motivate a retreat from the universalist standpoint. The universalist ideal of shared humanity with respect for dignity, human rights and equality is not dependent on the color of skin. It is the respect of fundamental values that makes the world human. To avoid corruption, the Enlightenment ideals should be "wrenched away from European hands and made the possession of all humanity" (Malik 2023, 170).

The beginning of the previous chapter refers to a world without color that has lost much of its attractive and pleasant qualities. Instead of eradicating color, a broader perspective of bioethics acknowledges that color is perceived in a range of nuances and that accepts that human life is colorful. Such perspective takes power and diversity as fundamental to critical analysis, and utilizes a theoretical framework that is genuinely intercultural and global, i.e. relevant for all people, ethnicities and cultures around the world. It is remarkable that the discipline of bioethics, that in the 1970s in Western countries replaced traditional medical ethics, is currently being transformed into a more inclusive approach that encompasses the globe. It presents a system of ethics that is worldwide in scope. This has become unavoidable since many of the ethical challenges in healthcare nowadays are global (e.g. pandemics, organ trade, malnutrition, migration and environmental degradation). These problems affect the whole of humankind, regardless of where people live, and they threaten not simply individual health and well-being but the health and survival of humanity. They also require global cooperation and action, necessitating a search for common ground, even when moral values in specific countries and regions will differ (Ten Have and Gordijn 2014). The global dimension of today's moral challenges also requires a broader ethical perspective. The confrontation with a new type of problem that is no longer localized in character but global in scope demands an approach that transcends the views and values of the Western culture in which bioethics originally emerged. Global bioethics in this sense not only refers to an expanded field of work but at the same time to an inclusive and comprehensive ethical orientation that departs from the usual emphasis on individual, medical and short-term perspectives.

Global bioethics as an encompassing, inclusive and planetary perspective is inspired by the ideals of cosmopolitanism: the unity of humanity, solidarity, equality, openness to differences and focus on what human beings have in common (Ten Have 2016). In this philosophy, human beings are considered as citizens of their own community, state (polis) or culture, as well as citizens of the world (cosmos). In the first, they are born and grown up; they share a common origin, language and customs with co-citizens. In the second, they participate because they belong to humanity; all human beings share the same dignity and equality. Being a citizen of the world liberates the individual from captivity in categories such as culture, tradition and community, but also gender and race. Cosmopolitanism acknowledges that human beings are connected to other beings and the surrounding world.

This anthropological experience of "connectedness" and "togetherness" is taken as the starting-point for global ethical reflection. If human beings not only interact with each other but also belong together and are mutually dependent, then relationships and shared responsibility in shaping the world play a defining role in who a person is. The notion of individual autonomy as used in mainstream bioethics should then be redefined as a relational concept. Community, mutual support, social responsibility, cooperation and solidarity should have a significant role in inclusive and comprehensive bioethical discourse. Furthermore, being situated in a web of connections is a precarious experience. Because their bodies position them in the world, human beings are exposed to the world and other persons, necessarily implying vulnerability.

Relationality is therefore the core notion of global bioethics. It is a more fundamental characteristic of being human than relatedness and connectedness. A human person is continuously engaging in relations, but this is often conceived from the viewpoint of the individual. The notion of relationality expresses that individuals not merely connect and interact with each other but belong together and are mutually dependent, taking responsibility and shaping their lives together. As integrated wholes of body and soul they are embedded within communities, and they exist in a web of relationships with other beings and the environing world. The first experience of humans is that the world is shared with others. Authentic human being is being-together.

Against the background of new global challenges and the need for a broad and inclusive approach, global bioethics works with a range of ethical principles. Without disregarding the value of personal autonomy and individual rights, it develops a moral discourse with a more extensive horizon. First, it argues that human beings are not abstract and de-contextualized individuals: they are necessarily embedded in social structures. Beneficial social, cultural, economic and political conditions make flourishing in health possible. This implies that power differences and structural violence should be critically addressed, and that principles such as justice and equity play a major role in bioethical debate. The concept of the common good is rehabilitated since individual persons are citizens concerned with shared interests that are not simply the aggregation of private interests. Furthermore, new forms of collective engagement and agency will be necessary to influence the systemic conditions that produce global problems. Over the past few decades, neoliberal policies have made life more precarious for most human beings (as well as for other living creatures). They have created a context of structural violence and multiplied opportunities for injustice and exploitation. Though individual actions and concerns are important, they will not be sufficient to bring about social transformation. The power structure of neoliberal globalization as the source of bioethical problems should be criticized with a broader set of moral concepts such as human vulnerability, social responsibility, equity, justice, sharing of benefits and future generations within bioethics discourse. These concepts will direct bioethical attention to structural determinants of health and disease rather than individual decisions concerning care.

Second, diversity has become a central concern in global bioethics. Previously, we have discussed how Western culture has tended towards chromophobia in its attitude towards colors. Colors are distrusted and reality is often presented as black and white, which are not regarded as colors themselves (Batchelor 2000). This traditional way of thinking—of moving black and white outside of the world of colors—seems to be reiterated in the current ideology of colorblindness. The idea that colors are better ignored also corresponds with the antirealist theory of the nature of colors: they are illusory, intrinsically subjective and only exist in the mind. Assuming that colors essentially reside in the human mind is coherent with the ethical priority of individualism. The individual subject can indeed act as if colors do not matter since they do not belong to the surrounding world but are our own product. In this book, I have argued that this view of colors does not recognize the phenomenological experience of color in human existence. It is theoretically unsatisfactory because it discounts the functional role of colors in our relationship to the world, and how colors express identities. People communicate through color; it is a language without words, evoking an impressive range of meanings, and conveying various values and ideas. Colors also make the world beautiful and good in an aesthetic as well as ethical sense. For artists such as Kandinsky, colors are not a medium between observer and object, but they are the atmosphere in which the observer dwells (Riley 1995).

The effect of chromophobia and colorblindness is a reduced view of diversity. In Western culture, for three centuries (from the 17th to the 20th century) black and white are considered as noncolors, and black is contrasted to white. However, for most of its history this has not been the case. At least until the Renaissance, in the West three colors are regarded as basic, and black and white are contrasted with red (Pastoureau 2009). Black has been the original color; the oldest pigments were probably black. Already in ancient times there were many blacks, with different degrees and qualities: light and dark, matte and glossy, intense and delicate. The same is true for white which has various shades and nuances.

Experiencing the rich variety of colors in our life-world should prompt us to re-evaluate the notion of diversity. Respect for cultural diversity and pluralism has nowadays been recognized as one of the principles of global bioethics (Ten Have 2017). The awareness that many moral challenges to health and healthcare today have a global dimension implies that they are no longer dependent on the specifics of a particular culture or society. While it remains important to address these challenges at the local, national and regional level, coping with them requires international and global cooperation, as the recent Covid-19 pandemic has illustrated. Such coping presupposes that at least some fundamental values are shared in order to formulate effective policies around the globe. The efforts and difficulties in doing so have been apparent in recent examples of the activities of international organizations, such as the World Health Organisation's response to the coronavirus disease. That there are similar bioethical problems in almost all countries does not imply that the same ethical assessment and approach is used everywhere. The least one can say is that the planetary dimension of health challenges necessitates a rethinking of our usual ethical frameworks. It makes us aware of the "locality" of our moral views, while at the same time encouraging the search for moral views that are shared globally. This implies recognition of the fact that the dominant bioethical approach, based on a limited set of principles, is a product of Western, White culture. If this approach is universalized and applied across the globe it will be an example of ethical imperialism. Nonetheless, this recognition does not imply that it is not possible to reach agreement on principles that can be used universally across borders. This has been the exact mission of UNESCO, adopting in 2005 general principles to guide decisions and practices in global bioethics (in the *Universal Declaration on Bioethics and Human Rights*). Formulating these principles was undertaken at the request of developing countries that wanted similar normative principles to be applied in healthcare and medical research in order to avoid unequal practices and double standards. Such principles should not reiterate and extend the Western individualistic perspective of ethics but take into consideration the value systems of other cultures and countries across the world. That a trans-cultural moral approach is possible has been asserted by non-Western scholars. Jing-Bao (2005) argued the importance of exploring non-Western cultures to uncover their advocacy for universal principles. It is a mistake to assume that such principles (e.g. human dignity) are alien to and incompatible with these cultures. It is *moral protectionism* to assume that ethical principles, even after having emerged and being formulated in Western culture, continue to remain the property of such specific culture, and therefore are not universal but only valid within this specific context. Cultures differ but this does not imply that common standards and universal principles do not exist. It is moreover a mistake to assume that universalism is abstract; values and ideals are derived from particular histories, traditions and places (Malik 2023).

Global bioethics, as it has emerged since the turn of the millennium, does not simply promote universal values or acknowledge moral diversity. It is a dialectical effort to bridge universalism and particularism. Its main challenge is to combine and bridge convergence and divergence of values, and it is therefore not a finished product. How can recognition of differences in moral views and approaches be reconciled with the convergence towards commonly shared values? Criticisms of global bioethics often presuppose simplistic views of globalization. While worldwide interconnectedness bridges the gap between distance and proximity, some scholars assume a radical contrast between moral strangers and friends, while others fear the growth of a bioethical monoculture (Ten Have 2016). But it is not correct that globalization produces either uniformity or multiplicity; it does both. Just as the concept of race is impossible to attribute to individuals, people nowadays are part of multiple cultures. It is not clear where their roots exactly are. They may consider themselves at the same time as Dutch, European and citizen of the world. The same is true for the notion of culture itself. No culture today is monolithic and pure. All cultural traditions are dynamic; they have changed and are changeable; they are necessarily a mélange of different components. Differences do not exclude that there is a common core. The term *interculturalism* is therefore more appropriate than multiculturalism since it acknowledges diversity while at the same time insisting on universal values. The term "interculturality" emphasizes interaction. "Inter" refers to separation but also linkage and communication. The supposition is that we can position ourselves between cultures; we can occupy a place between the universal and the particular. It means that we recognize similarities between self and other that can be the basis for dialogues between cultures, and at the same time that we can maintain differences and sustain boundaries between self and other. In other words, conceptually and practically, we are "in-between," moving beyond dualistic, binary thinking, adopting universalizing as well particularizing practices simultaneously (Lobo, Marotta and Oke 2011).

While multiculturalism emphasizes respect for diversity, individual freedom, justice and equal treatment, interculturalism introduces a moral vocabulary of interaction, dialogue, participation, trust, cooperation and solidarity. It is not sufficient to have multiple co-existent value systems and respect them; rather, the challenge is to produce and cultivate practices that can create community. If there is common ground, it needs to be cultivated through interaction and communication. Convergence is not a given but is rather the result of an ongoing activity of deliberation, consultation and negotiation. It is exactly this "interstitial perspective" that motivates the development of global bioethics.

6.10 Conclusion

It could not have been a real surprise that the Covid-19 pandemic disproportionately affected people of color. Health and healthcare disparities for these populations have already existed for a long a time. But the pandemic, in conjunction with the Black Lives Matter movement, was a wakeup call that placed issues such as racism, structural injustice, discrimination and vulnerability more center stage in bioethics. Since then, the pervasiveness of the moral associations of white and black (and to a lesser extent, other colors) as well as their deleterious effects on health and healthcare have become major topics of concern in ethical debate.

In previous chapters it is shown that in the history of medicine and healthcare, colors have played a significant, and generally positive role. They are regarded as diagnostic and prognostic clues about what is going on inside the human body; they are indicators of physiological and pathological processes; they suggest particular medicinal effects. Colors themselves are often interpreted as relaxing or exciting, and have long been used as remedies. The synthetic production of colors in the nineteenth century laid the foundations for the modern pharmaceutical industry. This symbiotic relationship between color and health collapsed as soon as color was connected to the idea of race. The discriminatory and classificatory functions of color are projected on human beings rather than on the surrounding world, and give rise to normative judgments of superiority or inferiority. In order to avoid these negative impacts, the prevailing view in policy and science nowadays is that color should not be noticed as a relevant issue. In the context of bioethics, the result of colorblindness is that issues such as race and racism are insufficiently addressed in bioethical discourse.

This chapter elaborates how bioethics should deal with color. It emphasizes that race should be taken seriously as an ethical problem. Given its negative implications, the concept of race should preferably be eliminated in healthcare and medical science. The challenge for bioethical analysis is to critically focus on settings in which the notion continues to be used: language, disease conditions, clinical practice, research and medical education. The second, and related challenge concerns racism. Bioethics cannot be silent about racism because racism evidently violates crucial ethical concerns and principles such as justice and human dignity. Racism is also a significant source of medical harm since it is a barrier to health and healthcare for people of color. Bioethics should furthermore acknowledge that racism is still pervasive in contemporary societies. Cultural changes, policies and legislation have not eradicated racism but have made it less openly and explicitly practiced. Racist attitudes, beliefs and behaviors are now generally regarded as aberrations and exceptions, manifested at the level of implicit prejudices and biases. While it is important to counter such biases, especially in the context of healthcare, and to "de-program" the often unintended normative associations of colors, racism also persists because it is incorporated in institutional and organizational practices. Such systemic racism is invisible. Examples mentioned in this chapter are clinical guidelines and algorithms, research findings, and the use of so-called "normal" values which almost automatically put people of color at a disadvantage, since most of the data is related to White subjects. The implication of persistent racism is that bioethical analysis must not only focus on the perspective of individual patients and healthcare providers but should address the contextual and structural dimensions of health and disease, and of the systems and services that are supposed to care for all people.

Confronting and interrogating race and racism demand a

reorientation of bioethics. The core argument of this chapter is that bioethics should be "chromophilic"—not blind to colors or affiliated with only one color, white or black. This calls for a review of the methods as well as contents of bioethical discourse. Ethics used to be regarded as a rational undertaking of argumentation and deliberation. Recently, however, the role of emotions and feelings in moral judgments has been reassessed. Most moral judgments are not the outcome of reasoning but primarily made as soon as a situation, condition or act is perceived as morally relevant, and based on emotions triggering an intuitive response. Only then is moral reasoning applied to justify the judgment to other people. In this view, bioethics is like aesthetics, and should not only focus on rational arguments but also on the intuitive stage in which moral judgments are immediately and automatically delivered. In regard to notions of race and racism, this opens up the way for bioethics to go beyond the rational level of ethical principles that evidently condemn these notions, and to direct its critical attention to the emotional level, where color associations and normative evaluations arise in inconspicuous ways.. Bioethical analysis can help to identify why and how such associations emerge, and are reinforced in language and imagery. Most of all, bioethical analysis can use moral imagination to make individuals aware of the perspective of other people, and to better understand the experiences of others.

Besides employing other methods, a race-conscious bioethics should also redefine its contents. Ongoing experiences with racism illustrate the shortcomings of the ideology of individualism which permeates bioethics as well as healthcare, social and economic policies. People of color are discriminated against and stigmatized because they belong to racialized groups or categories that are systematically disadvantaged. Whereas the effects are harmful and disrespectful for individuals, the roots of the problem are at a different level: the dehumanizing system of unjust social relations which is the result of historic legacies of oppression and exploitation. If bioethical discourse wants to address issues of race and racism it therefore needs to concentrate its critical attention at this systemic level. This chapter argues that two concepts in particular are important: power and diversity. Additionally, the chapter highlighted how an inclusive and enlarged conception of bioethics has emerged that provides more intellectual and moral tools to scrutinize power differences and diversity concerns. This so-called global bioethics applies a range of ethical principles, which not only articulate individualistic values (such as personal autonomy) but also communal and social values (such as vulnerability, justice and solidarity) as well as environmental ones (e.g. respect for biodiversity, and future generations). Operating with a range of principles, global bioethics shows that it acknowledges diversity and, at the same time, aspires to determine the values human beings and various cultures share and have in common. It confirms that the world is full of colors, enjoyable, beautiful and valuable.

References

Akinlade, O. 2020. Taking black pain seriously. *New England Journal of Medicine* 383 (10): e68 (1–2).

Amutah, C., Greenidge, K., and Mante, A. et al. 2021. Misrepresenting race—The role of medical schools in propagating physician bias. *New England Journal of Medicine* 384 (9): 872–878, https://doi.org/10.1056/nejmms2025768

Anderson, E. 2010. *The imperative of integration*. Princeton, NJ and Oxford: Princeton University Press.

Andorno, R., and Pele, A. 2016. Human dignity. In: ten Have, H. A. M. J. (ed.), *Encyclopedia of global bioethics*. Cham: Springer Nature, 1537–1546.

Bachelard, G. 2014. *On poetic imagination and reverie*. New York: Spring Publications.

Baker, R. 2003. Balkanizing bioethics. *The American Journal of Bioethics* 3 (2): 13-14.

Barilan, Y. M., and Brusa, M. 2013. Deliberation at the hub of medical education: Beyond virtue ethics and codes of practice. *Medicine, Health Care and Philosophy* 16: 3–12, https://doi.org/10.1007/s11019-012-9419-3

Batchelor, D. 2000. *Chromophobia*. London: Reaktion Books.

Berg, S. 2023. AMA: Use of BMI alone is an imperfect clinical measure. *AMA*, 14 June, https://www.ama-assn.org/delivering-care/public-health/ama-use-bmi-alone-imperfect-clinical-measure

Bhopal, R., and Donaldson, L. 1998. White, European, Western, Caucasian, or what? Inappropriate labelling in research on race, ethnicity, and health. *American Journal of Public Health* 88: 1303–1307.

Bieri, P. 2017. *Human dignity: A way of living*. Cambridge: Polity Press.

Blair, I. V., Ma, J. E., and Lenton, A. P. 2001. Imagining stereotypes away: the moderation of implicit stereotypes through mental imagery. *Journal of Personality and Social Psychology* 81 (5): 828–841.

Blum, L. 2002. *I'm not a racist, but... The moral quandary of race*. Ithaca, NY: Cornell University Press.

Bonilla-Silva, E. 2022. *Racism without racists. Color-blind racism and the persistence of racial inequality in America*. Lanham: Rowman & Littlefield.

Bradby, H. 2012. Race, ethnicity and health: The costs and benefits of conceptualising racism and ethnicity. *Social Science & Medicine* 75: 955–958, https://doi.org/10.1016/j.socscimed.2012.03.008

Brandt, A. M. 1978. Racism and research: The case of the Tuskegee Syphilis Study. *Hastings Center Report* 8(6): 21–29.

Braun, L., Fausto-Sterling, A., Fullwiley, D., Hammonds, E. M., Nelson, A. et al. 2007. Racial categories in medical practice: How useful are they? *PLoS Medicine* 4 (9): e271, https://doi.org/10.1371/journal.pmed.0040271

Braun, L. 2014. *Breathing race into the machine. The surprising career of the spirometer from plantation to genetics*. Minneapolis, MN and London: University of Minnesota Press, https://doi.org/10.5749/minnesota/9780816683574.001.0001

Braun, L., and Saunders, B. 2017. Avoiding racial essentialism in medical science curricula. *AMA Journal of Ethics* 19 (6): 518–527, https://doi.org/10.1001/journalofethics.2017.19.6.peer1-1706

Braun, L., Wentz, A., Baker, R., Richardson, E., and Tsai, J. 2021. Racialized algorithms for kidney function: Erasing social experience. *Social Science & Medicine* 268:113548, https://doi.org/10.1016/j.socscimed.2020.113548

Brown, M. K., Carnoy, M., Currie, E. et al. 2023. *Whitewashing race. The myth of a color-blind society*. Oakland, CA: University of California Press, https://doi.org/10.1525/9780520394605

Bueno Pimenta, F. J., and Garcia Gomez, A. 2023. Contemplating the principles of the UNESCO declaration on bioethics and human rights: A bioaesthetic experience. *International Journal of Ethics Education* 8 (2): 249–274, https://doi.org/10.1007/s40889-023-00176-8

Burchard, E. G., Ziv, E., Coyle, N. et al. 2003. The importance of race and ethnic background in biomedical research and clinical practice. *New England Journal of Medicine* 348 (12): 1170–1175.

Burgess, D., van Ryn, M., Dovidio, J., and Saha, S. 2007. Reducing racial bias among health care providers: Lessons from social-cognitive psychology. *Journal of General Internal Medicine* 22 (6): 882–887.

Callier, S. L. 2019. The use of racial categories in precision medicine research. *Ethnicity & Disease* 29 (Suppl 3): 651-658, https://doi.org/10.18865/ed.29.s3.651

Christakis, N. A. 2019. *Blueprint. The evolutionary origins of a good society*. New York, Boston, London: Little, Brown Spark.

Constantino, J. N., Abbacchi, A. M., Saulnier, C. et al. 2020. Timing of the diagnosis of autism in African American children. *Pediatrics*. 146 (3): e20193629, https://doi.org/10.1542/peds2019-3629

Crenner, C. 2014. Race and laboratory norms. The critical insights of Julian Herman Lewis (1891–1989). *Isis* 105 (3): 477–507, https://doi.org/10.1086/678168

Council of Europe. 2024. *ECRI (European Commission against Racism and Intolerance)*, http:/www.coe.int/ecri

Derkse, W. 1992. *On simplicity and elegance. An essay in intellectual history*. Delft: Eburon.

Dessie, A. S., and Chen, E. H. 2023. Reporting race and ethnicity in medical research: You can't fix what you don't know. *European Journal of Emergency Medicine* 30: 157–158, https://doi.org/10.1097/mej.0000000000001026

Devine, P. G., Forscher, P. S., Austin, A. J., and Cox, W. T. 2012. Long-term reduction in implicit race bias: A prejudice habit-breaking intervention. *Journal of Experimental Social Psychology* 48 (6): 1267–1278, https://doi.org/10.1016/j.jesp.2012.06.003

Elias, A., and Ben, J. 2023. Pandemic racism: Lessons on the nature, structures, and trajectories of racism during COVID-19. *Bioethical Inquiry* 20: 617–623, https://doi.org/10.1007/s11673-023-10312-0

European Commission, Migration and Home Affairs. 2024. Racism. *European Commission*, https://home-affairs.ec.europa.eu/networks/european-migration-network-emn/emn-asylum-and-migration-glossary/glossary/racism_en

Fanon, F. 2021. *Black skin, white masks*. London: Penguin Books (original 1952).

Fesmire, S. 2003. *John Dewey and moral imagination. Pragmatism in ethics*. Bloomington and Indianapolis, IN: Indiana University Press.

Flanagan, A., Frey, T., and Christiansen, S. L. 2021. Updated guidance on the reporting of race and ethnicity in medical and science journals. *Journal of the American Medical Association* 326 (7): 621–627, https://doi.org/10.1001/jama.2021.13304

Fombonne, E., and Zuckerman, K.E. 2022. Clinical profiles of Black and White children referred for autism diagnosis. *Journal of Autism and Developmental Disorders* 52: 1120–1130, https://doi.org/10.1007/s10803-021-05019-3

Fredrickson, G. M. 2002. *Racism. A short history*. Princeton, NJ and Oxford: Princeton University Press.

Fullilove, M.T. 1998. Comment: abandoning "race" as a variable in public health research—an idea whose time has come. *American Journal of Public Health* 88 (9): 1297–1298.

Fullwiley, D. 2007. Race and genetics: Attempts to define the relationship. *BioSocieties* 2 (2): 221–237, https://doi.org/10.1017/s1745855207005625

Futterman, J., Bi, C., Crow, B. et al. 2024. Medical educators' perceptions of race in clinical practice. *BMC Medical Education* 24: 230, https://doi.org/10.1186/s12909-024-05232-5

Gerrits, E. M., Assen, L. S., Noordegraaf-Eelens, L., Bredenoord, A. L., and van Mill, M. H. W. 2023. Moral imagination as an instrument for ethics education for biomedical researchers. *International Journal of Ethics Education* 8 (2): 275–289, https://doi.org/10.1007/s40889-023-00171-z

Gombault, C., Grenet, G., Segurel, L. et al. 2023. Population designations in biomedical research: Limitations and perspectives. *HLA* 101 (1): 3–15, https://doi.org/10.1111/tan.14852

Gould, S. J. 1996. *The mismeasure of man*. New York and London: W. W. Norton & Company.

Gov.UK. 2015. *Equality Act 2010: Guidance*, https://www.gov.uk/guidance/equality-act-2010-guidance

Gutin, I. 2019. Essential(ist) medicine: Promoting social explanations for racial variation in biomedical research. *Medical Humanities* 45: 224–234, https://doi.org/10.1136/medhum-2017-011432

Haidt, J. 2001. The emotional dog and its rational tail: A social intuitionist approach to moral judgment. *Psychological Review* 198 (4): 814–834.

Howard-Jones, N. 1979. On the diagnostic term "Down's disease." *Medical History* 23: 102–104.

Hunter, M. 2007. The persistent problem of colorism: Skin tone, status, and inequality. *Sociology Compass* 1 (1): 237–254, https://doi.org/10.1111/j.1751-9020.2007.00006.x

Hyry-Beihammer, E. K., Lassila, E. T., Estola, E., and Uitto, M. 2022. Moral imagination in student teachers' written stories on an ethical dilemma. *European Journal of Teacher Education* 45 (3): 435–449, https://doi.org/10.1080/02619768.2020.1860013

Ioannidis, J. P. A., Powe, N. R., and Yancy, C. 2021. Recalibrating the use of race in medical research. *Journal of the American Medical Association* 325 (7): 623–624, https://doi.org/10.1001/jama.2021.0003

Jablonski, N. G. 2012. *Living color. The biological and social meaning of skin color*. Berkeley, CA and London: University of California Press.

Jablonski, N. G. 2021. Skin color and race. *American Journal of Physical Anthropology* 175: 437–447, https://doi.org/10.1002/ajpa.24200

Jalali, Y., Matheis, C., and Lohani, V. K. 2022. Imagination and moral deliberation: A case study of an ethics discussion session. *International Journal of Engineering Education* 38 (3): 709–718.

James, J. E., and Iacopetti, C. L. 2021. Beyond seeing race: Centering racism and acknowledging agency within bioethics. *The American Journal of Bioethics* 21 (2): 56–58, https://doi.org/10.1080/15265161.2020.1861380

Jantzen, D., Newton, L., Dompierre, K-A., and Sturgill, S. 2023. Promoting moral imagination in nursing education: Imagining and performing. *Nursing Philosophy*: e12427, https://doi.org/10.1111/nup.12427

Jing-Bao, N. 2005. Cultural values embodying universal norms: A critique of a popular assumption about cultures and human rights. *Developing World Bioethics* 5 (3): 251–257.

Johnson, M. 1993. *Moral imagination. Implications of cognitive science for ethics*. Chicago and London: The University of Chicago Press.

Johnstone, M.-J., and Kanitsaki, O. 2010. The neglect of racism as an ethical issue in health care. *Journal of Immigrant and Minority Health* 12: 489–495, https://doi.org/10.1007/s10903-008-9210-y

Kaplan, J. B., and Bennett, T. 2003. Use of race and ethnicity in biomedical publication. *Journal of the American Medical Association* 289: 2709–2716, https://doi.org/10.1001/jama.289.20.2709

Keane, W. 2016. *Ethical life. Its natural and social histories*. Princeton, NJ and Oxford: Princeton University Press.

Lamont, M. 2023. *Seeing others. How recognition works—and how it can heal a divided world*. New York: One Signal Publishers/Atria.

Lobo, M., Marotta, V., and Oke, N. (eds) 2011. *Intercultural relations in a global world*. Champaign, IL: Common Ground Publishing.

Lorusso, L., and Bacchini, F. 2023. The indispensability of race in medicine. *Theoretical Medicine and Bioethics* 44: 421–434, https://doi.org/10.1007/s11017-023-09622-6

Macneill, P. 2017. Balancing bioethics by sensing the aesthetic. *Bioethics* 31: 631–643, https://doi.org/10.1111/bioe.12390

Malik, K. 2023. *Not so black and white. A history of race from white supremacy to identity politics*. London: Hurst & Company.

Malinowska, J. K., and Żuradzki, T. 2023. Reductionist methodology and the ambiguity of the categories of race and ethnicity in biomedical research: An explanatory study of recent evidence. *Medicine, Health Care and Philosophy* 26: 55–68, https://doi.org/10.1007/s11019-022-10122-y

Matthew, D. B. 2015. *Just medicine. A cure for racial inequality in American health care*. New York and London: New York University Press.

Mesman, J. 2021. *Opgroeien in kleur. Opvoeden zonder vooroordelen.* [Growing up in color. Educating without prejudices] Amsterdam: Uitgeverij Balans.

Metzl, J. M. 2009. *The protest psychosis: How schizophrenia became a Black disease.* Boston, MA: Beacon Press.

Metzl, J. M., and Hansen, H. 2014. Structural competency: Theorizing a new medical engagement with stigma and inequality. *Social Science & Medicine* 103: 126–133, https://doi.org/10.1016/j.socscimed.2013.06.032

Ministerie van Binnenlandse Zaken en Koninkrijksrelaties. 2022. *Juridische definitie van racism.* Brief aan de Voorzitter van de Tweede Kamer der Staten Generaal. Kenmerk 2022-0000327692.

Minow, M. 2022. Restorative justice and anti-racism. *Nevada Law Journal* 22 (3): 1157–1178, https://doi.org/10.2139/ssrn.4298677

Morning, A. 2008. Ethnic classification in global perspective: A cross-national survey of the 2000 census board. *Population Research and Policy Review* 27 (2): 239–272, https://doi.org/10.1007/s11113-007-9062-5

Molewijk, A. C., Abma, T., Stolper, M., and Widdershoven, G. 2008. Teaching ethics in the clinic. The theory of moral care deliberation. *Journal of Medical Ethics* 34: 120–124.

Moscou, S., Anderson, M. R., Kaplan, J. B., and Valencia, L. 2003. Validity of racial/ethnic classifications in medical records data: An exploratory study. *American Journal of Public Health* 93 (7): 1084–1086.

Neal, R.E. and Morse, M. 2021. Racial health inequities and clinical algorithms. A time for action. *Clinical Journal of the American Society of Nephrology* 16 (7): 1120-1121.

Neville, H. A., Awad, G. H., Brooks, J. E., Flores, M. P., and Bluemel, J. 2013. Color-blind racial ideology. Theory, training, and measurement implications in psychology. *American Psychologist* 68 (6): 455–466, https://doi.org/10.1037/a0033282

Nieblas-Bedolla, E., Christophers, B., Nkinsi, N. T. et al. 2020. Changing how race is portrayed in medical education: Recommendations from medical students. *Academic Medicine* 95 (12): 1802–1806. https://doi.org/10.1097/acm.0000000000003496

Noë, A. 2023. *The entanglement. How art and philosophy make us what we are.* Princeton, NJ and Oxford: Princeton University Press, https://doi.org/10.1515/9780691239293

Parikh, R. B., Teeple, S., and Navathe, A. S. 2019. Addressing bias in artificial intelligence in health care. *Journal of the American Medical Association* 322 (24): 2377–2378, https://doi.org/10.1001/jama.2019.18058

Pastoureau, M. 2009. *Black. The history of a color.* Princeton, NJ and Oxford: Princeton University Press.

Pastoureau, M. 2014. *Green. The history of a color*. Princeton, NJ and Oxford; Princeton University Press.

Pastoureau, M. 2019. *Yellow. The history of a color*. Princeton, NJ and Oxford: Princeton University Press.

Ratti, E., Russo, F. 2024. Science and values: A two-way direction. *European Journal for Philosophy of Science* 14 (6), https://doi.org/10.1007/s13194-024-00567-8

Ray, K. 2023. *Black health. The social, political, and cultural determinants of black people's health*. New York: Oxford University Press, https://doi.org/10.1093/oso/9780197620267.001.0001

Richie, C. 2018. *Principles of green bioethics*. East Lansing: Michigan State University Press.

Riley, C. A. 1995. *Color codes. Modern theories of color in philosophy, painting and architecture, literature, music and psychology*. Hanover and London: University Press of New England.

Roberts, D. 2011. *Fatal invention. How science, politics, and big business re-create race in the twenty-first century*. New York and London: The New Press.

Rodríguez-Hernández, M. L., and Montoya, E. 2011. Fifty years of evolution of the term Down's syndrome. *Lancet* 378 (9789): 402, https://doi.org/10.1016/s0140-6736(11)61212-9

Rubin, R. 2021. Tackling the misconception that Cystic Fibrosis is a "White People's Disease." *Journal of the American Medical Association* 325 (23): 2330–2332, https://doi.org/10.1001/jama.2021.5086

Rudman, L. A., Ashmore, R. D., and Gary, M. L. 2001. "Unlearning" automatic biases: The malleability of implicit prejudice and stereotypes. *Journal of Personality and Social Psychology* 81 (5): 856–868, https://doi.org/10.1037//0022-3514.81.5.856

Russell, C. 2022. Meeting the moment: Bioethics in the time of Black Lives Matter. *American Journal of Bioethics* 22 (3): 9–21, https://doi.org/10.1080/15265161.2021.2001093

Saini, A. 2019. *Superior. The return of race science*. Boston, MA: Beacon Press.

Shanawani, H., Dame, L., Schwartz, D. A., and Cook-Deegan, R. 2006. Non-reporting and inconsistent reporting of race and ethnicity in articles that claim associations among genotype, outcome, and race or ethnicity. *Journal of Medical Ethics* 32: 724–728, https://doi.org/10.1136/jme.2005.014456

Shelby, T. 2014. Racism, moralism, and social criticism. *Du Bois Review* 11 (1): 57–74, https://doi.org/10.1017/s1742058x14000010

Smith-McLallen, A., Johnson, B. T., Dovidio, J. F., and Pearson, A. R. 2006. Black and white: The role of color bias in implicit race bias. *Social Cognition* 24 (1): 46–73, https://doi.org/10.1521/soco.2006.24.1.46

Steinkamp, N., and Gordijn, B. 2003. Ethical case liberation on the ward. A comparison of four methods. *Medicine, Health Care and Philosophy* 6: 235–246.

Stift, R. 2024. Jaren wachten op een diagnose: want autisme is toch iets van witte mensen? [Waiting years for a diagnosis since autism is an affection of white people?] *NRC*, 29 March.

Ten Have, H. 2016. *Global bioethics. An Introduction*. London and New York: Routledge.

Ten Have, H. 2017. Respect for cultural diversity and pluralism. In: Tham, J., Kwan, K. M., and Garcia, A. (eds), *Religious perspectives on bioethics and human rights*. Cham: Springer International Publishing, 3–23.

Ten Have, H. ed. 2018. *Global education in bioethics*. Cham: Springer Nature.

Ten Have, H. A. M. J. 2019. *Wounded planet. How declining biodiversity endangers health and how bioethics can help*. Baltimore, MD: Johns Hopkins University Press.

Ten Have, H., and Gordijn, B. 2014. Global bioethics. In: ten Have, H. A. M. J., and Gordijn, B. (eds), *Handbook of global bioethics*. Dordrecht: Springer Publishers, 3–18.

Ten Have, H., and Pegoraro, R. 2022. *Bioethics, healthcare and the soul*. London: Routledge.

Tsai, J., Ucik, L., Baldwin, N. et al. 2016. Race matters? Examining and rethinking race portrayal in preclinical medical education. *Academic Medicine* 91 (7): 916–919, https://doi.org/10.1097/acm.0000000000001232

Tsai, J, Cerdeña, J. P., Khazanchi, R., Lindo. E., Marcelin, J. R., Rajagopalan, A., Sandoval, R. S., Westby, A., and Gravlee, C. C. 2020. There is no "African American physiology": The fallacy of racial essentialism. *Journal of Internal Medicine* 288 (3): 368–370, https://doi.org/10.1111/joim.13153

UNESCO. 1978. Declaration on race and racial prejudice. *OHCHR*, https://www.ohchr.org/en/instruments-mechanisms/instruments/declaration-race-and-racial-prejudice

University of Bristol. 2024. Black and Brown in bioethics: BBB 2024 conference. *Black and Brown in Bioethics*, 20 April, https://blackbrownbioethics.blogs.bristol.ac.uk/2024/04/20/conference-engaging-diversity-in-bioethics-theory-and-practice/

Varcoe, C., Browne, A. J., Wong, S., and Smye, V. L. 2009. Harms and benefits: Collecting ethnicity data in a clinical context. *Social Science & Medicine* 68 (9): 1659–1666, https://doi.org/10.1016/j.socscimed.2009.02.034

Vyas, D. A., Eisenstein, L. G., and Jones, D. S. 2020. Hidden in plain sight—Reconsidering the use of race correction in clinical algorithms. *New England Journal of Medicine* 383: 874–882, https://doi.org/10.1056/nejmms2004740

Warnock, G. J. 1971. *The object of morality*. London: Methuen & Co Ltd.

Xue, W., and White, A. 2021. Covid-19 and the rebiologisation of racial difference. *The Lancet* 398: 1479–1480, https://doi.org/10.1016/s0140-6736(21)02241-8

Yearby, R. 2021. Race based medicine, colorblind disease: How racism in medicine harms us all. *The American Journal of Bioethics* 21 (2): 19–27, https://doi.org/10.1080/15265161.2020.1851811

Yudell, M., Roberts, D., DeSalle, R., and Tishkoff, S. 2016. Taking race out of human genetics. *Science* 351 (6273): 564–565, https://doi.org/10.1126/science.aac4951

Zack, N. 2023. *Ethics and race. Past and present intersections and controversies.* Lanham: Rowman & Littlefield.

Index

achromatopsia 112
Addison's disease 85
aesthetics 3, 13, 22–24, 47, 67–68, 72, 82, 111–112, 147, 181–182, 184–187, 192, 200, 205
Africa 20, 124, 126, 128, 141, 166
Agassiz, Louis 132
albinism 5, 85, 145
alchemy 98, 128
algorithms 19, 163, 168, 204
alizarin 80, 98, 100
alkaloids 97
Allen, Theodore 129
American Medical Association 141, 168
Amsterdam 93
analgesis 90, 100
anatomy 11, 35, 80, 84, 86, 88, 105–106, 132–133
Ancient Egypt 59, 92–94, 122
Ancient Greece 8, 29, 55, 122, 127, 131
Ancient Rome 8, 60, 114, 116, 121, 127, 188
Anderson, E. 196
aniline 80, 88, 96–97, 99–100
animal world 113
Anubis 122
apothecaries 80, 90, 93, 98
Arab countries 166
Aristotle 10, 31–32, 49
arsenic 82, 94–95, 100, 189
Asia 5, 61, 93, 120, 124, 126–127, 137, 145, 159, 164–165
aspirin 100

ater 5
Australia 159
autism 167
automatic thinking 190
Avicenna 91, 102

Bachelard, Gaston 190–191
Bacon, Francis 93
bacteriology 80, 99–100
Badische Anilin und Soda Fabrik (BASF) 97
basic color terms 49, 51–52, 71
Batchelor, David 17, 120
Bayer, Friedrich 97, 100
Benedictines 115
Berlin, B. 49, 51–52
Bernier, Francois 123–125, 128, 130
bias 137–139, 143, 148, 165, 168, 170–171, 173, 175, 179, 188–190, 194, 196, 204
bioethics 3, 10, 17–18, 21–22, 24, 34, 50, 131, 141–145, 148–149, 160–161, 164, 166, 168, 171, 173–174, 177–181, 185–187, 189–191, 193, 195, 197–205
  agenda of 21
  and aesthetics 3, 22, 24, 182, 187, 205
  and diversity 146, 180, 197, 205
  and race 144, 146, 148, 164, 168, 171. *See also* bioethics: colorblindness of; *See also* race: moral problems of
  and racism 173–174, 177
  and relationality 198
  as monochromatic 146
  as whiteness 146, 180

balkanization of 180
colorblindness of 21, 142, 149, 178–179, 204
color of 144
emergence of 179
global 180, 198, 201–202, 206
green 180
interstitial perspective of 203
principles of 22, 143–144, 161, 179, 181, 195, 199, 201, 205–206
biomedical research 164, 168–169, 171
Birren, F. 65
black 3–8, 11, 13, 15, 17–20, 31–32, 35, 48, 51–52, 59–60, 62–64, 79, 83, 85, 90, 93, 111–112, 114, 116, 118–125, 127–131, 137, 139, 143, 145–149, 159, 161–163, 166–168, 174, 177, 180, 187–189, 200, 203, 205
anthropological theories of 121, 123, 126
as degeneration 126, 130
as pollution 8
as skin color 7, 125, 129–131, 137, 139, 159, 161–163, 166–168
associations of 8, 63–64, 115, 120, 122–123
meaning of 6, 123
Black Death 5, 116
Black diseases 159, 163
Black Lives Matter 18, 203
blue 1–3, 5–6, 8, 10, 13–17, 29, 32, 35, 42, 48–52, 54–56, 59–62, 65–66, 68–69, 79, 81, 86–88, 90, 92, 97, 100–104, 112, 114, 117, 128, 188
and codes 6–7
and disease 5–7, 86
and meaning 60
and performance 13–14, 50, 66
associations of 1, 13–15, 63, 91, 188
blue baby syndrome 5
Blumenbach, Johann Friedrich 18, 125–126, 128–131, 133–134, 165, 183
Blum, L. 134–136, 160
body mass index (BMI) 168

books 95
Boyle, Robert 87, 93
Broca, Paul 132
brown 3–4, 7, 17–18, 52, 60, 63, 79, 86, 88, 90, 93, 117, 126, 132, 179–180
Buddhist tradition 61
Bueno Pimenta, F. J. 184

camouflage 39, 113
cancer 4, 80, 83, 85, 88–89, 107
cardiac catheterization 88
carmine 88
Caucasian 18, 126, 161, 164–165, 183
Cave of Altamira 58
celare 9, 29, 118
Celsus, Aulus Cornelius 83–84, 91
Celts 127, 129
Cézanne, Paul 93
chemistry 47, 80, 96, 98, 111
chemotherapy 100, 103, 107
China 61, 102, 114, 188
Chirimuuta, M. 36–37, 40, 44
chlorosis 6
cholera 106, 122
Christian culture 62
chromoclasm 8
chromophobia 9, 117–119, 188, 200
chromosomes 97
chromotherapy 16, 56, 81, 102
Ciba 97
Cicero 114
Cistercians 115
classical Antiquity 17, 60, 117
Clore, G. L. 121
Clusius, Carolus 93
coal tar 96–97, 99–100
colonialism 113, 124, 133, 141, 176, 197
color
and bioethics 142
and clothing 8, 18, 40, 47, 59, 66, 116, 119
and disease. *See* Black diseases; *See* blue: and disease; *See* green: and disease; *See* pink disease

and emotion 3, 6, 9, 13–14, 17, 48–49, 54, 57, 63–64, 66, 72, 81, 112, 117, 123, 143, 147, 181
and food 14, 40, 69
and healthcare facilities 16, 81, 104
and interior design 16, 62, 69, 81, 104
and language 1, 13, 29, 48–50, 52, 58, 70, 200. *See also* Sapir-Whorf hypothesis
and light 1, 10–11, 32, 35–36, 48, 59, 68, 72
and marketing 14–15, 47, 50, 68, 72, 92
and meaning 5–7, 13–15, 31, 44, 53, 56, 58, 61, 63, 67–68, 71, 92, 114, 121, 187
and medicine 80, 90–92
and neurophysiology 11, 35–36, 53
and normativity 13, 17, 24, 47, 72, 113–114, 118–119, 123, 129, 131, 143, 147, 181, 187, 189, 204–205
and pain 15
and performance 13, 49, 66–68, 72
and personality 10, 56, 71
and race. *See* black: as skin color; *See* white: as skin color; *See* race: as skin color
and racism. *See* white: as privilege; *See* white: as skin color; *See* black: as skin color
and rationality 118, 187
and societal learning 67
and temperature 55, 112
and therapy 15, 24, 71
and wavelength 7, 16, 30, 37, 102, 104
as a secondary quality 9, 34
as danger 7
as disposition 37–38, 40
as exotic 17
as experience 43
as illusion 24, 30, 35–36, 118
as perception 35–36, 50, 63, 68, 72, 112
as pharmakon 31
as signage 104
classification of 51, 55, 128
nature of 10, 29, 34, 40, 47–48, 51, 53, 67, 149, 200
philosophies of 34, 43
polarity of 56
power of 15, 24, 50, 56, 70–71, 92, 104, 149
spectral 10, 32, 35, 119, 128
warm and cold 55–56, 64, 69
color ascetism 118
colorblindness 10, 21, 112, 142, 146, 148–149, 173, 178–179, 181, 194, 196, 200, 204
color codes 6, 18, 59, 116, 118
color cure 79, 101
colores austeri 8, 114
colores floridi 8, 114
color-in-context theory 67, 72
colorism 19, 21, 149, 183
color names 48, 70
color theories 30, 41, 60, 62, 71, 113
  adverbialism 41
  antirealism 35–36, 40, 43, 51, 53, 200
  ecological theory 12, 38–40, 43, 113, 125
  phenomenological theory 12, 43, 51, 144, 200
  realism 30, 36, 40
  relationism 30, 40, 43–44, 51, 62, 67, 144, 181
color therapy 81, 102
coronavirus. *See* Covid-19 pandemic
cosmology 4
cosmopolitanism 198
Council of Europe 172
Covid-19 pandemic 6–7, 18, 21, 112, 145, 159, 170, 201, 203
craniometry 131, 133
Cult of the Virgin 60
cultural diversity 197, 201
Curie, Marie and Pierre 95
cyanosis 3, 86
cystic fibrosis 167

Dalton, John 111–112
Dass, Angelica 20, 145
Democritus 31
dentistry 88
dermatitis 85
Descartes, René 33–35
desert locusts 39, 113
Dewey, John 193
diabetes 86, 88
diagnosis 15, 40, 79, 83–84, 88–89, 145, 160, 167–168, 171, 188
discrimination 7, 18, 21, 40, 129, 133, 135–138, 140, 145, 148, 159, 172–174, 178–180, 183, 195–196, 203
Domagk, Gerhard 100
Down, John Langdon 165
Down's syndrome 165
dress codes 116, 187
Dutch National Institute for Public Health and the Environment (RIVM) 70
Dyer, R. 144

ecology 60, 62
Egyptian papyrus 92
Ehrlich, Paul 88, 99–100, 107
electromagnetic spectrum 51, 65
Elliot, A. J. 67
emerald 93, 95
England 91, 106, 116
Enlightenment 60, 133–134, 177, 197
environmental justice 175
equality 161, 175, 177, 197–198
erysipelas 5
ethical imperialism 201
ethics 3, 6–7, 17–18, 21–24, 60, 72, 115, 133, 138, 141–143, 146, 148, 160–161, 169, 171, 173–174, 178–182, 184–187, 190–193, 195–206. *See also* bioethics
Europe 1, 5, 9, 17, 20, 60–61, 93, 102, 106, 113, 117, 119, 124–125, 128, 133, 169, 172, 189
European Commission 172
European Parliament 172
European Union 7–8

facies Hippocratica 3, 83
Fanon, Frantz 197
fashion industry 101
Fehling's test 80
Finsen, Niels 104
food 14, 38, 40, 50, 62, 69, 72, 93, 112, 125, 162, 174
France 8, 101, 116, 135, 169

Galen 79, 83, 91, 106, 122, 127
gamboge 5, 93–94
Garcia Gomez, A. 184
Gassendi, Pierre 125
Geigy 97
Germany 64, 97, 101, 135, 169
glaucoma 5
Gobineau, Arthur de 134
Goethe, Wolfgang von 9, 54–57, 64, 71, 111, 120
Goldstein, Kurt 64–65
Golgi, Camillo 88
Gould, S. J. 134–135
Gram, Hans Christian 89
Greek sculptures 121, 131
green 5–6, 10, 13–17, 30, 32, 35, 38, 42, 48–49, 51–52, 55–56, 59–60, 62, 64–65, 68–69, 80–83, 90, 93–95, 101, 103–105, 111–112, 114–115, 117, 128, 180, 188
  and disease 5. *See also* green disease
  and hospitals. *See* hospital green
  and meaning 6, 59, 70
  and toxicity 60, 70, 95, 189
  associations of 13, 15, 62–63, 189
  radioactive 95
  Scheele's 94
green disease 5–6, 82
Greenland 62
grey 13, 17, 52, 60, 63, 68, 82, 111–112, 115, 117

Hades 122
Hardin, C. L. 52
Hardy, Françoise 1
Harvey, William 86

health disparities 139–140, 162, 170, 174
Hippocrates 3, 79, 83, 91, 127
histology 80, 88, 100
HIV 86
Hoechst 100
Hofmann, August Wilhelm 96–97
Holland 98. *See also* Netherlands
hospital green 16, 105
hospitals 6, 16, 69, 79, 81, 91, 104–105
human dignity 161, 177–178, 201, 204
Hume, David 36, 131
humors 4, 84, 102, 127
hygiene 16–17, 60–61, 86, 90, 100–101, 120–122, 147, 189

immigrants 19, 21, 129, 134, 147, 149, 167
India 97, 102, 124, 166
indigo 9–10, 32, 35, 60, 88, 97, 113
Indonesia 120
inequality 133, 138, 142, 172, 174, 177–178, 196
injustice 18, 22, 140, 142, 146, 148–149, 175–176, 180, 196, 199, 203
Institute of Medicine 139
interculturalism 197, 202–203
interior design 16, 62, 68, 72, 81, 104
International Association of Color Consultants/Designers 47
intuitive judgments 24, 186–187
Inuits 49–52
Ireland 21, 129
Islamic civilizations 59, 188–189
Italy 79

Jablonski, N. G. 131
Jantzen, D. 193
Japan 58–59, 102, 120
jaundice 3–4, 16, 81, 83, 103–104
Jefferson, Thomas 130–131
Jung, Carl Gustav 56

Kandinsky, Wassily 13, 48, 200
Kant, Immanuel 9, 130, 132–133, 177, 183

Kay, P. 49, 51–52
khroma 29
kidney function 168
Koch, Robert 100, 107
kohl 93

lavender 90
leadership 56
lead white 29, 94
Leclerc, George-Louis, count of Buffon 125
Leonardo da Vinci 49
leukemia 5
Levinas, Emmanuel 23
life-world 12, 24, 31, 41–42, 44, 53, 113, 200
linguistic relativity 49–52, 71
Linnaeus, Carolus 125, 127, 129–130
Lister, Joseph 100
litmus test 15, 80, 86–87
Locke, John 33–34, 37, 125
Lombroso, Cesare 132
Lucretius 31
Lüscher, Max 56–57

madder 88, 97
magenta 97
magic bullet 100
Maier, M. A. 67
Malevich, Kazimir 58
marketing 14–15, 47, 50, 68, 72, 92
Matisse, Henri 13, 48
mauveine 88, 96, 106
medical education 170–171, 193, 204
medical language 171
medication 5, 15–16, 29, 62, 70, 80, 86, 90–94, 96–97, 99, 101, 103, 105–106, 137, 163
melancholy 1, 4–5, 56, 79, 84, 127–128
melanoma 5
melasma 85
mercury 82, 98
Merleau-Ponty, Maurice 12, 41–42
Mexico 64
microbes 80
Middle Ages 8, 60, 84, 90, 117, 128, 188

Middle East 124
minerals 29, 80, 91–92, 94, 98
mongolism 165
moral deliberation 23, 185–187, 190, 193
moral experience 22, 182, 186–187, 190, 193
moral imagination 182, 190, 192–195, 205
moral judgment 23, 182, 185–187, 189–190, 205
moral perception 23, 182, 193, 195
moral protectionism 202
moral purity 8, 121
moral reasoning 23, 182, 185–187, 189–190, 193–195, 205
moral theology 17, 114
Morning, A. 169
Morton, Samuel 131–132
multiculturalism 202–203
mummy brown 93
Munsell Color System 63–64
Myser, C. 143–144

Naples yellow 94
Napoleon 95
National Institutes of Health 168
National Insurance Bill 91
Netherlands 87, 135, 145, 167, 176
newborns 16, 104
Newton, Isaac 10–11, 31–32, 35, 49, 59–60, 119, 128
New Zealand 159
Nexium 92
North America 60

ochronosis 5
Olympic Games 66
ophthalmology 88
orange 3, 6, 10, 13, 15, 32, 40, 52, 55, 88, 90, 93, 98, 102
osteoporosis 168
oxazepam 91

Painter, Nell Irvin 129
Pakistan 87
Paleolithic caves 122
pallor 3, 82
Papanicolaou, George 89
Paracelsus 98
Pastoureau, Michel 17, 62, 90, 113
pathology 6, 79, 83–84, 86, 88, 106, 170
perception 3, 11–12, 14–15, 23–24, 33–38, 40–43, 49–53, 55, 62–63, 66, 68–70, 72, 81, 101, 104–105, 111–113, 166, 182, 186
Perkin, William 96–97, 99, 106
pharmaceutical industry 5, 203
pharmacology 62, 101
pharmacy 62, 90, 93
phenol 99
phototherapy 81, 104
physicalism 11, 32, 34, 36, 38
physician's dress 122
Picts 113
pigments 5, 9, 17, 29, 35, 40, 60–61, 71, 80, 86–88, 91–96, 98, 100–101, 106, 113–114, 118–119, 121–122, 189, 200
  and medication 80, 91, 93–94
  and painting 2, 5, 8–9, 13, 47–48, 57–58, 70, 92, 94, 101, 106, 118, 122, 185–187
  artificial 80, 97–99, 106–107, 121
  natural 29, 80, 93, 121
  toxicity 94
pink 52, 63, 66, 82, 85
pink disease 82
placebo drugs 90
plague 5, 102, 116
plantations 9, 113
Plato 31, 146–147
Plinius 8, 49, 93, 114
pluralism 146, 201
poison 29, 39, 94–95, 189
Poland 64
pollution 8, 121–122
Ponza, G. L. 79, 101

porphyria 4, 81
power differences 160, 174–175, 196, 199, 205
principle of justice 174
prison cells 66
Prontosil 100
Provence 98
Prussian blue 60
psoriasis 85, 104
public health 7, 99, 141
purple 3, 52, 81, 89, 92, 97, 114, 116

quinine 96–97

race 7, 10, 18–19, 24, 120, 123–145, 147–148, 159–179, 182–183, 188–190, 194–198, 202–205
  and bioethics. See bioethics: and race
  and disease 166–167
  and UNESCO 134
  as a social construct 135
  as fiction 134
  as skin color 18, 123, 125–129, 131, 161–162
  classification of 18, 123–125, 127–129, 131, 133–134, 161–162, 164, 169
  concept of 123, 125, 131, 134, 160, 204
  in biomedical research 168–171
  in clinical settings 167
  in medical education 170
  moral problems of 160, 162
race correction 167
race norming 19, 168
racial groups 160–161, 163, 166–167, 170, 177
racial science 131–133, 143, 147–148
racism 7, 18–19, 21, 24, 125, 127, 132, 134–138, 140–142, 145, 147–148, 159–160, 163, 168–169, 172–175, 177–179, 181, 195–196, 203–205
  and bioethics. See bioethics: and racism
  and healthcare 137, 141–142, 159, 163, 173–174, 196, 204
  as a threat to public health 141
  as practice 19, 135–138, 141–142, 145, 148, 163, 167–168, 172–173, 179, 195, 204
  consequences of 163
  explaining 138
  institutional 196
  interpersonal 138, 162, 176
  systemic 136, 138, 140, 142, 146, 162–163, 166, 170, 173, 175–176, 204
radiation 20, 95, 104
red 4–8, 10, 13–16, 18, 20, 31–32, 35, 37, 40, 48–49, 51–52, 54–56, 58–60, 62, 64–69, 79, 81–82, 84–85, 87–90, 92, 97–98, 100–102, 104, 111–112, 114, 122, 126, 128, 132, 143, 145, 189, 200
  and competition 14, 56, 66
  and intellectual performance 13, 49, 66–67
  and pain 16
  and tips 14, 68
  associations of 8, 13–15, 59, 63
  as warning signal 6, 59, 67
redness 3, 37, 82–83, 85
Reformation 8, 17, 117, 188
relationality 12, 14, 24, 37–38, 40–41, 43, 48, 66–67, 143–144, 178, 198–199
Rembrandt 8, 119
Renaissance 200
respect for personal autonomy 161, 179
retina 11, 35, 88, 104–105
Roberts, D. 162
Romantic era 55–56, 60, 71, 189
Rome 60, 95. See also Ancient Rome
rubella 5, 81
Russell, C. 18, 143, 174
Russia 64

Sacks, O. 112
saffron 88, 93
Saini, A. 135
Saint Bernard of Clairvaux 118
Salvarsan 100, 107
Sandoz 97

sanitarian movement 99
Sapir-Whorf hypothesis 49–50
scarlet fever 5, 86
Scheele, Carl 94–95
schizophrenia 167
Schopenhauer, Arthur 36
Second World War 172
Seneca 8, 114
Sherman, G. D. 121
sickle-cell anemia 166
sin 59, 115, 121–122
skin bleaching 183
slavery 9, 127, 129, 133, 136, 140, 148, 169, 176
smallpox 16, 81, 102, 104
smoking 70, 86
social determinants of health 139, 171
solidarity 196, 198, 203, 206
Soviet Union 169
Spain 169
spinach green 16, 105
staining techniques 88
stigmatization 147, 169, 172, 194, 196, 205
stool 4, 83, 104
Strabo 94
sub-Saharan Africa 20, 130
sumptuary laws 116
sumptuary Laws 116
surgery 100, 163
sustainability 60, 62, 189
Switzerland 97
synthetic dyes 5, 80, 88, 98, 100–101, 103
syphilis 100, 107, 141, 167

Tanguy, Julien 93
Texas 129
tongue 85
tuberculosis 20, 82, 100, 107, 166
Turner, William 57
Tuskegee syphilis study 141, 167

UNESCO 134, 201
United Kingdom 6, 137, 145, 169, 172

United Nations 8, 87, 135–136, 145, 176–177
United States 2, 7, 19, 63–64, 66, 102, 129, 135, 137, 140, 145, 147, 159, 166–167, 169, 174, 176
Universal Declaration of Human Rights 177
Universal Declaration on Bioethics and Human Rights 184, 201
universalism 53, 197, 202
University of Bristol 180
urine 3, 80–81, 83–85, 88, 104
uroscopy 79, 84, 106, 188

van Gogh, Vincent 93–94
van Leeuwenhoek, Antonie 88
Vermeer, Johannes 92
vermillion 98
Vesalius, Andreas 86
Victorian culture 121, 188
violet 10, 32, 51, 79, 103
vitamin B12 86
vitamin D 21
vitiligo 85
von Baeyer, Adolf 97
vulnerability 7, 18–19, 137, 144, 159, 175, 180, 195–196, 198–199, 203, 206
Vyas, D. A. 168

Wasserman, R. 112
wavelengths 1–2, 7, 11, 16, 30, 32–37, 65, 102, 104
weather maps 112
Werther 56
Western civilization 8, 19, 24, 59, 128, 147, 161, 198, 200, 202
white 3, 5, 7–8, 10–11, 13–21, 29, 31–32, 34–37, 48–49, 51–52, 59–64, 66, 68–69, 82, 84–86, 88, 90–91, 94, 101, 105, 111–112, 114, 117–134, 137, 140, 142–149, 159, 162–168, 174, 176–177, 179–180, 183, 187–189, 200–201, 203–205
  as moral purity 8, 121
  as privilege 19, 129, 142, 148, 176, 180

as rationality 129, 147
as skin color 7, 18–20, 29, 126, 128, 130, 132–133, 140, 143–145, 148–149, 162–163, 167–168, 180, 183
associations of 8, 10, 17, 61, 64, 90, 105, 115, 120, 122
synthetic 94
White Plague 82
White supremacy 18, 144, 148
Whorf, B. L. 51. *See* also Sapir-Whorf hypothesis
Winckelmann, Johann 131
wokeism 19, 149
World Health Organization (WHO) 139

yellow 3–5, 8, 10, 13, 15, 17–18, 20, 32, 35, 39, 48, 51–52, 54–56, 59–60, 63–65, 79–81, 83–84, 90, 93–94, 103, 105, 114, 117, 126, 128, 132, 143, 145, 188
as poison 39, 94
associations of 8, 13, 15, 61, 63–64, 188
stigmatizing 114
Yellow Emperor 188
yellow fever 5

About the Team

Alessandra Tosi was the managing editor for this book.

Adèle Kreager proof-read the final version; Annie Hine compiled the index; Laura Rodriguez created the Alt-text.

Jeevanjot Kaur Nagpal designed the cover. The cover was produced in InDesign using the Fontin font.

Jeremy Bowman typeset the book in InDesign and produced the paperback, hardback and EPUB editions. The main text font is Tex Gyre Pagella and the heading font is Californian FB.

Cameron Craig produced the PDF and HTML editions. The conversion was performed with open-source software and other tools freely available on our GitHub page at https://github.com/OpenBookPublishers.

Raegan Allen was in charge of marketing.

This book was peer-reviewed by an anonymous referee. Experts in their field, our readers give their time freely to help ensure the academic rigour of our books. We are grateful for their generous and invaluable contributions.

This book need not end here...

Share

All our books — including the one you have just read — are free to access online so that students, researchers and members of the public who can't afford a printed edition will have access to the same ideas. This title will be accessed online by hundreds of readers each month across the globe: why not share the link so that someone you know is one of them?

This book and additional content is available at
https://doi.org/10.11647/OBP.0443

Donate

Open Book Publishers is an award-winning, scholar-led, not-for-profit press making knowledge freely available one book at a time. We don't charge authors to publish with us: instead, our work is supported by our library members and by donations from people who believe that research shouldn't be locked behind paywalls.

Join the effort to free knowledge by supporting us at
https://www.openbookpublishers.com/support-us

We invite you to connect with us on our socials!

BLUESKY	MASTODON	LINKEDIN
@openbookpublish .bsky.social	@OpenBookPublish @hcommons.social	open-book-publishers

Read more at the Open Book Publishers Blog
https://blogs.openbookpublishers.com

You may also be interested in:

Chance Encounters
A Bioethics for a Damaged Planet
Kristien Hens (author), Christina Stadlbauer, Bart H.M. Vandeput (illustrators)

https://doi.org/10.11647/OBP.0320

Bioethics
A Coursebook
COMPOST Collective

https://doi.org/10.11647/OBP.0449

Beyond Price
Essays on Birth and Death
J. David Velleman

https://doi.org/10.11647/OBP.0061

Towards an Ethics of Autism
A Philosophical Exploration
Kristien Hens

https://doi.org/10.11647/OBP.0261

www.ingramcontent.com/pod-product-compliance
Lightning Source LLC
Chambersburg PA
CBHW061254230426
43665CB00027B/2943